Revolutions in Differential Equations

Exploring ODEs with Modern Technology

Library of Congress Catalog Card Number 99-62786
ISBN 0-88385-160-1
Printed in the United States of America
Current Printing (last digit):
10 9 8 7 6 5 4 3 2 1

Revolutions in Differential Equations

Exploring ODEs with Modern Technology

Edited by

Michael J. Kallaher

Published and distributed by
The Mathematical Association of America

The MAA Notes Series, started in 1982, addresses a broad range of topics and themes of interest to all who are involved with undergraduate mathematics. The volumes in this series are readable, informative, and useful, and help the mathematical community keep up with developments of importance to mathematics.

MAA Notes

11. Keys to Improved Instruction by Teaching Assistants and Part-Time Instructors, *Committee on Teaching Assistants and Part-Time Instructors, Bettye Anne Case,* Editor.
13. Reshaping College Mathematics, *Committee on the Undergraduate Program in Mathematics, Lynn A. Steen,* Editor.
14. Mathematical Writing, by *Donald E. Knuth, Tracy Larrabee, and Paul M. Roberts.*
16. Using Writing to Teach Mathematics, *Andrew Sterrett,* Editor.
17. Priming the Calculus Pump: Innovations and Resources, *Committee on Calculus Reform and the First Two Years,* a subcomittee of the Committee on the Undergraduate Program in Mathematics, *Thomas W. Tucker,* Editor.
18. Models for Undergraduate Research in Mathematics, *Lester Senechal,* Editor.
19. Visualization in Teaching and Learning Mathematics, *Committee on Computers in Mathematics Education, Steve Cunningham and Walter S. Zimmermann,* Editors.
20. The Laboratory Approach to Teaching Calculus, *L. Carl Leinbach et al.,* Editors.
21. Perspectives on Contemporary Statistics, *David C. Hoaglin and David S. Moore,* Editors.
22. Heeding the Call for Change: Suggestions for Curricular Action, *Lynn A. Steen,* Editor.
24. Symbolic Computation in Undergraduate Mathematics Education, *Zaven A. Karian,* Editor.
25. The Concept of Function: Aspects of Epistemology and Pedagogy, *Guershon Harel and Ed Dubinsky,* Editors.
26. Statistics for the Twenty-First Century, *Florence and Sheldon Gordon,* Editors.
27. Resources for Calculus Collection, Volume 1: Learning by Discovery: A Lab Manual for Calculus, *Anita E. Solow,* Editor.
28. Resources for Calculus Collection, Volume 2: Calculus Problems for a New Century, *Robert Fraga,* Editor.
29. Resources for Calculus Collection, Volume 3: Applications of Calculus, *Philip Straffin,* Editor.
30. Resources for Calculus Collection, Volume 4: Problems for Student Investigation, *Michael B. Jackson and John R. Ramsay,* Editors.
31. Resources for Calculus Collection, Volume 5: Readings for Calculus, *Underwood Dudley,* Editor.
32. Essays in Humanistic Mathematics, *Alvin White,* Editor.
33. Research Issues in Undergraduate Mathematics Learning: Preliminary Analyses and Results, *James J. Kaput and Ed Dubinsky,* Editors.
34. In Eves' Circles, *Joby Milo Anthony,* Editor.
35. You're the Professor, What Next? Ideas and Resources for Preparing College Teachers, *The Committee on Preparation for College Teaching, Bettye Anne Case,* Editor.
36. Preparing for a New Calculus: Conference Proceedings, *Anita E. Solow,* Editor.
37. A Practical Guide to Cooperative Learning in Collegiate Mathematics, *Nancy L. Hagelgans, Barbara E. Reynolds, SDS, Keith Schwingendorf, Draga Vidakovic, Ed Dubinsky, Mazen Shahin, G. Joseph Wimbish, Jr.*
38. Models That Work: Case Studies in Effective Undergraduate Mathematics Programs, *Alan C. Tucker,* Editor.
39. Calculus: The Dynamics of Change, *CUPM Subcommittee on Calculus Reform and the First Two Years, A. Wayne Roberts,* Editor.
40. Vita Mathematica: Historical Research and Integration with Teaching, *Ronald Calinger,* Editor.
41. Geometry Turned On: Dynamic Software in Learning, Teaching, and Research, *James R. King and Doris Schattschneider,* Editors.

42. Resources for Teaching Linear Algebra, *David Carlson, Charles R. Johnson, David C. Lay, A. Duane Porter, Ann E. Watkins, William Watkins,* Editors.
43. Student Assessment in Calculus: A Report of the NSF Working Group on Assessment in Calculus, *Alan Schoenfeld,* Editor.
44. Readings in Cooperative Learning for Undergraduate Mathematics, *Ed Dubinsky, David Mathews, and Barbara E. Reynolds,* Editors.
45. Confronting the Core Curriculum: Considering Change in the Undergraduate Mathematics Major, *John A. Dossey,* Editor.
46. Women in Mathematics: Scaling the Heights, *Deborah Nolan,* Editor.
47. Exemplary Programs in Introductory College Mathematics: Innovative Programs Using Technology, *Susan Lenker,* Editor.
48. Writing in the Teaching and Learning of Mathematics, *John Meier and Thomas Rishel.*
49. Assessment Practices in Undergraduate Mathematics, *Bonnie Gold,* Editor.
50. Revolutions in Differential Equations: Exploring ODEs with Modern Technology, *Michael J. Kallaher,* Editor.

These volumes can be ordered from:
MAA Service Center
P.O. Box 91112
Washington, DC 20090-1112
800-331-1MAA FAX: 301-206-9789

Introduction

The teaching of differential equations is presently undergoing a vast change. This change is based on two factors. First, one outcome of the "Reform Calculus" movement was the introduction and heavy use of computers and computer experiments in the basic calculus courses. Students now entering the beginning Differential Equations course have solid experience in using computers and computer programs (including graphics and computer algebra programs) to explore and solve mathematics problems. Second, because of the tremendous advances in computers, the investigation of differential equations has turned again to the qualitative side.

The qualitative study of differential equations goes back to Poincare at the end of the last century. It has continued, but only at the research and applied level. One tool that played an important role in this study was the analog computer. However, for several reasons it was not feasible to emphasize the qualitative analysis in the beginning differential equations courses. Instead, the algebraic solution of equations was the main thrust of such courses resulting in the memorization of formulas by students with little or no understanding of the mathematics or the underlying processes being modeled by the equations. In many colleges the basic differential equations course was the archetypical cookbook course. The course mainly covered solution formulas and some clever tricks, many applications from science and engineering, suitably sanitized for both teacher and student, and maybe a little exposure to numerical methods.[1]

Over the last ten years change has begun occurring, helped along by the availability of computing equipment and mathematical software—e.g., Mathematica and Maple—for students. Today, software (e.g., IPE, ODE Architect) especially for the differential equations course is becoming available. All of this means that today differential equations can be taught from the qualitative point of view with emphasis on the underlying mathematics and physical processes that give rise to the equations.

A positive effect of the new computer hardware and software is the ability to explore differential equations and dynamical systems more deeply and completely. A second outcome is the use of realistic models for explaining the mathematics of the course.

The articles in this volume deal with the teaching of a modern course in differential equations. They are intended for the teacher, not the student. That is, the articles do not constitute the equivalent of a textbook; rather, a teacher should study and reflect on them with a view toward applying ideas relevant to the teacher's particular circumstances.

A central theme is the incorporation of modern technology into the differential equations course. Technology is changing rapidly today, so rapidly that it will have an enormous effect on the course, even possibly to the point of eliminating the need for the classroom. Both internal pressures (those from within the mathematical community) and external pressures for change will be discussed.

The combined effect of technology and reform calculus has caused many to re-examine the teaching of differential equations—both content and pedagogy. Many textbooks recently published are thus very different from each other and from previous editions, both in approach and in content. It is a tremendous task to work carefully through all of these texts to see what is available. This volume provides one efficient way to learn about the direction in which the content and teaching of differential equations is going.

The authors represent the spectrum of the field of differential equations. They have performed first-class research into the solution and understanding of such equations—in closed form, numerically, and graphically—and they are experienced teachers of ODE courses who have individually and collectively accumulated expert knowledge on the

[1]On a personal level the editor in 1960 had a differential equations course out of the text *A Short Course in Differential Equations* by E.D. Rainville. A comparison with recent texts—for example, *Differential Equations* by R. Borrelli and C. Coleman, *Differential Equations: A Dynamical Systems Approach, Part I* by J. Hubbard and B. West, or *Exploring Differential Equations via Graphics and Data* by D. Lomen and D. Lovelock—shows how much the differential equations course is being changed.

incorporation of modern technology into the ODE class. Several have experience in the development of computer software either for solving ODEs or for use in the ODE class. These experiences give them unique insights into the future of the ODE class.

Borrelli and Coleman, authors of textbooks and laboratory manuals for ODE courses, discuss several ODE solvers suitable for the first ODE course. They also discuss and give examples—from engineering, agriculture, chemistry—of modeling exercises that take advantage of such solvers. They are careful to point out important points to consider when using a laboratory in the ODE course.

Boyce looks at the effects modern technology will have on the teacher, the syllabus, and pedagogy in the differential equation classroom. Changes in the content and method of instruction are discussed: topics to be eliminated or given less coverage, topics to be introduced or expanded in coverage. Boyce argues strongly for replacing the classical lecture with the studio mode of instruction, which involves a laboratory setting and emphasis on individual teacher-student interaction. His view of the future is provocative and needs careful consideration.

Branton and Hale discuss the geometric background and view of differential equations and systems (linear and nonlinear). Using models of harvesting, electrical circuits, and spring-mass systems they explain the geometry of several concepts including stability, limit cycles and bifurcation.

Cooper and LoFaro review the Internet and the highly probable effect it and other technological developments of the last five years will have on the teaching of differential equations. These developments are exerting strong external pressures to change the teaching of most mathematical courses. By the use of three examples, Cooper and LoFaro discuss how the Internet can easily bring into the classroom large volumes of information which can quickly be assimilated into meaningful exercises illuminating the concepts of the course.

Lomen discusses the use of data sets to develop differential equations, suggests appropriate mathematical models, and lists methods for checking the applicability of such models. These sets may be obtained from texts, research articles, experiments from science or engineering laboratories, from the Internet, or from student experiments. Examples are given to illustrate several ways for using the data in a differential equations course.

Manoranjan gives an intuitive and elementary introduction to dynamical systems within the context of differential equations. The treatment starts with basic concepts and brings the reader to the point where nonlinear differential equations can be understood and appreciated. At the informal level he discusses the basic topics and gives an introduction to investigative techniques frequently used in applying differential equations to solve real-world problems.

In recent years numerical methods for solving differential equations have come to play an important role in ODE courses. Shampine and Gladwell argue persuasively that in discussing numerical methods the emphasis should be on what students must know in order to use quality software for gaining insight about solutions and their behavior. After they briefly present the fundamentals of numerical methods from this perspective, they give an overview of quality software available for initial value problems in several areas of application, including comments on how the area can influence the techniques used. Pointers to software for solving boundary value problems are also provided.

West discusses the benefits of technology in teaching differential equations. Having been involved since the early 1980's in adapting technology for the classroom, she points out the advantages of visualization, interactive computer graphics, and other computer attributes for helping the teacher to encourage student thinking about and understanding of the concepts and techniques developed in the differential equations course.

As is evident in the above brief descriptions, these articles do overlap somewhat. However, just as in the classroom understanding is helped by approaching concepts from more than one side, different viewpoints on the uses of technology will help the teacher to develop his/her own appreciation for and ability to adapt technology for the classroom. It is our hope that the reader will be convinced that modern technology is an important tool for helping the student understand and be comfortable with mathematical concepts and techniques.

I wish to thank the authors for taking time out of their busy schedules to think about and write on the impact modern technology is having and will have on the differential equations course. Many thanks are also due Jaimie Dahl, who deciphered our writing and translated it into readable text. Thanks also to Leona Ding, Donna Pierce, and Dave Richards for computerizing many of the figures for publication.

Michael J. Kallaher
July 1999

Contents

Modeling and Visualization in the Introductory ODE Course

Robert L. Borrelli
Courtney S. Coleman
Harvey Mudd College

Our increasingly technological society has welcomed computers with open arms. No longer the domain of the esoteric few, powerful and (relatively) inexpensive platforms are being marketed as a panacea for all that ails the civilized world. Software packages are being churned out by armies of programmers whose genius has produced code that not only addresses current societal needs but also opens doors for more technological change. Curiously, academia is the slowest segment of society to be affected by these changes. There are good reasons for this: Most science and engineering courses in the undergraduate curriculum have been around so long that their content and method of approach are pretty well defined, and the supporting textbooks reflect this fact. It's in the nature of academia to be slow to change—we aren't trying to find a niche in a competitive market. Nevertheless, there are some very encouraging signs over the past decade that academia is responding to the challenge of using computers in the curriculum in an effective and creative way.

Reform movements in calculus, linear algebra and differential equations are well under way, and all of them make good use of hands-on projects in connection with the modeling and visualization capabilities that technology provides. However these reform movements ultimately turn out in the twenty-first century, one thing is already clear: modeling and visualization will be in these courses in one way or another for a long time to come.

What is modeling, anyway? Scientists and engineers generally use the term to describe the process of translating a natural system into a form called a model that can be dealt with in a way that we have confidence in. The models that interest us here are mathematical models which involve ordinary differential equations.

Curricular reform these days seems to be edging closer and closer to an interdisciplinary approach in which students take a more active role in their own course work. In mathematics departments this approach usually involves laboratory-based activities in which students play a hands-on role in converting "word problems" into mathematical models and "solving" them. We will describe modeling activities which are suitable for a course in differential equations.

1. The First Course in Differential Equations

Trends in the undergraduate math curriculum have been influenced by a variety of factors. First and foremost is the ready availability of powerful platforms and excellent software for both numerical and symbolic computations. The reform movements in all disciplines have been affected by these technology twins, but the undergraduate mathematics curriculum has arguably been affected the most. The primary use of technology in instruction is to visualize mathematical concepts, solutions of equations, etc., in the context of a laboratory environment. The first course in ODEs seems to be an ideal place to use computers for modeling and visualization.

We have put in more years than we care to remember in redesigning the first course in ODEs at Harvey Mudd College to bring in modeling and visualization as an essential component of the course. When we first embarked on this path in the late '70's, we couldn't find any solvers which were both robust and easy-to-use, or any affordable platforms on which the solvers could run. So our first task was to design our own solver package that could be used reliably on available platforms in a dynamical systems lab environment. With the collaboration of four of our students (Ned Freed, Dan Newman, Kevin Carosso and Tony Leneis), we finally, after considerable effort, produced an ODE solver package which was suitable for our needs, and the National Science Foundation helped us to set up a lab dedicated to this course. At last we were in business, or so we thought.

As is often the case at universities and colleges with engineering programs, our first course in ODEs is a required 3-credit-hour sophomore course with no time set aside for laboratory instruction. The traditional syllabus for the course was designed to serve the needs of our client disciplines. There was little flexibility in the course syllabus and no chance the college would give us an additional credit-hour to set up a companion laboratory course, so what to do? Well, first we made our solver absolutely transparent to use (no mean feat) so that students could use it on their own from the get-go with very little instruction required. Our solver is network accessible, and so we included drivers for most graphics platforms. That way students could access the solver from their dorm rooms (or anywhere really) and use their favorite platforms as a front end: X-windows, PCs, MACs, etc. Printing on any academic laser printer could also be done over the network. Graphs are automatically imprinted with the user name, the date, and the time so that they can be retrieved at any time. Next, we created a collection of computer experiments (later published as a *DE Lab Workbook*) which were more-or-less self-contained and designed to go along with a standard ODE course. We assigned one (or more) experiments per week from this collection that the students could do as "homework". In fact, each experiment replaces one of the homework sets that otherwise would have been done that week. This slight syllabus change was necessary because students would be quick to see that we were cramming a four-credit hour course into three-credit hours.

Students use any solver they like for laboratory assignments. At the first lecture we distribute a hand-out describing all the currently available ODE solvers with instructions on access from the various college computer labs. This arrangement works well for a required course that anyone in the department may be assigned to teach: individual instructors are free to select any textbook/solver combination.

In contrast to the reform calculus experience, there has been little doubt about the direction in which the introductory ODE course should change; the only details to be worked out concern the resources which support different modes of instruction. From 1992 to 1997 the NSF-funded Consortium for ODE Experiments (CODEE) has published a newsletter with information about incorporating hands-on projects ("experiments") in an ODE course. Experiments always involve a modeling component and are designed to address some questions in the modeling environment.

2. ODE Solvers

There are a number of excellent ODE software packages that support modeling and visualization in a first course in ODEs. Some are commercially distributed, and some are shareware. Some are built to run as a component of a large multi-purpose package, and some are standalone ODE packages. The stand-alone ODE packages are usually designed to run on only one platform, whereas the multi-purpose packages run on all platforms. Visit the web site http://www.hmc.edu/codee/solvers.html for an overview of many of the available solvers (it's difficult to keep current on this).

Here are brief descriptions of three solver packages that differ in their approach, but are designed to provide a lab experience to go along with an ODE course.

Interactive Differential Equations (IDE): A collection of interactive tools designed to explore a single concept or application in an ODE course, such as the logistic equation, direction fields, oscillators, numerical methods, the phase plane, eigenvectors, Laplace transforms, series solutions, chaos, bifurcations, and many other topics, each with linked animations of the systems being modeled. An important feature of the package is that the tools demonstrate visually the connections between real world phenomena and the mathematical models that describe them.

There are 97 interactive illustrations, called tools, arranged into 31 lab collections. The intuitive point-and-click interface allows the user to interact with the tools by setting initial conditions and using sliders to vary parameters for both linear and nonlinear models. Each set of tools is accompanied by a workbook lab consisting of background, instructions, and experiments, with space for writing answers. An instructor's manual is available with sample answers filled in.

Although IDE does *not* include an open-ended solver that allows you to input any ODE, it delivers visually striking classroom demonstrations, and has proven invaluable for independent homework assignment, or full blown laboratory experiences.

These tools were developed by a team of mathematics faculty including John Cantwell, Jean Marie McDill, Steven Strogatz, and Beverly West; the software designer, who originated the package, was Hubert Hohn. The package is now available for Windows 95 as well as for the Macintosh.

For more information visit the Addison-Wesley Web site http://awi.aw.com or awi-info@aw.com, or phone Liz O'Neil at (617) 944-3700, ext. 2380.

Internet Differential Equations Activities (IDEA): This product can be thought of as an interactive virtual lab book for differential equations at the undergraduate level. IDEA has the basic goal of developing and disseminating software for numerical explorations of mathematical models using differential equations. These materials are available over the Web and can be used by anyone with connections to the Internet. The IDEA developers, Tom LoFaro and Kevin Cooper, have created tools to assist instructors in the development of their own Web materials and/or to contribute to the IDEA site. Currently, most explorations are based on biology, chemistry, and ecology. The IDEA approach provides resources that can give students an appreciation for research projects involving differential equations not found in traditional texts.

For more information visit IDEA at http://www.sci.wsu.edu/idea or contact lofaro@ trout.math.wsu.edu.

ODE Architect: With NSF/DUE support, the CODEE consortium, John Wiley & Sons, and the software house Intellipro are producing an interactive multimedia ODE package which is built over a robust solver engine designed by Larry Shampine. Scheduled for release in 1998, the CD-ROM runs on a PC under Windows 3.1 or better and is accompanied by a lab workbook of experiments. With video, sound, animation, and dynamic graphics, this interactive package provides motivation for modeling, analysis, visualization, discovery, and interpretation in any ODE lab environment. In addition to the solver tool there are 13 interactive modeling modules on ODEs and dynamical systems, as well as a library of ODEs with their dynamically generated solution curves and orbits. Each module leads the user through a model building process via several exploration screens, and ends up with questions. These questions take the user to the solver tool and to the accompanying lab workbook where the user is asked to carry out graphics-based experiments to explain what is going on. Users can enter their own ODEs and explore dynamical systems with 2D or 3D graphics or numerical tables by seeing what happens when data and parameters change. Graphs are editable and axes can be scaled and labeled, equidistant-in-time orbital points marked. Graphs of solutions and orbits can also be colorized, animated, displayed in various line styles, overlayed with graphs of functions, and graphed together with solution curves of other ODEs. All of this is possible with no programming or special commands to remember. ODE Architect has a report writing feature; graphs can be cut and pasted into reports.

For more information contact Barbara Holland at bholland@wiley.com.

3. Laboratory Experiments for an ODE Course

We can use technology to examine dynamical processes and their ODE models that would have been inaccessible only a few years ago. Hands-on experience in the setting of an ODE laboratory is at the heart of this approach. The "laboratory" might be a room with computers or individuals at home or in a dormitory working with their own computers, or even lecture demos. Whatever the mode, here are some central ideas that come up in the dynamical systems laboratory.

Modeling

- Derivative as rate of change
- Balance law: net rate of change = rate in − rate out
- Compartmental models
- Newton's force laws
- Circuit laws for voltages and currents
- Chemical law of Mass Action

Visualization

- Derivative as slope of a curve
- The art of making graphs that tell a story
- Effective use of computer graphics; choosing appropriate displays
- How to interpret graphs; extracting information from graphs
- Change scales, time span, viewpoint to get the most informative graph

Solution Behavior

- Do we use theory, formulas, or computer simulation to study solution behavior?
- Do solutions tend to an equilibrium state or a periodic solution with advancing time? Or do they become chaotic?
- How sensitive are solutions to changes in data and system parameters?
- Is analysis of the sign of a rate function useful to see how solutions behave?
- What happens to long-term solution behavior as a parameter changes?
- If the modeling system is nonlinear, can it be approximated by a linear system, or is the behavior due to the nonlinearities?

Computer Techniques

- Should the variables be scaled before computing?
- Scale out system parameters not relevant to the sensitivity study.
- Does your solver handle on-off functions of engineering, or can you work around it?
- Is the solution behavior generated by your numerical solver really there, or is it an artifact of the numerics? Has your solver overlooked an important aspect of solution behavior because its internal settings are inappropriate?

Well, that's a long list of things to keep in mind, but the numerical experiments that follow touch on many of these points. Don't look for the standard experiments and models that can be found in many places. Our aim is to present challenging models that can now be handled by students in an introductory ODE course if they have access to a decent numerical solver.

4. Will the Message Get Through?

What kind of message can be transmitted by a communications channel (such as a coaxial cable) and still be recognizable at the other end? Let's suppose our message is the voltage $\mathrm{V_{in}}$ for the circuit sketched below. Using Ohm's and Kirchhoff's Laws to find voltage drops in the loop, we see that

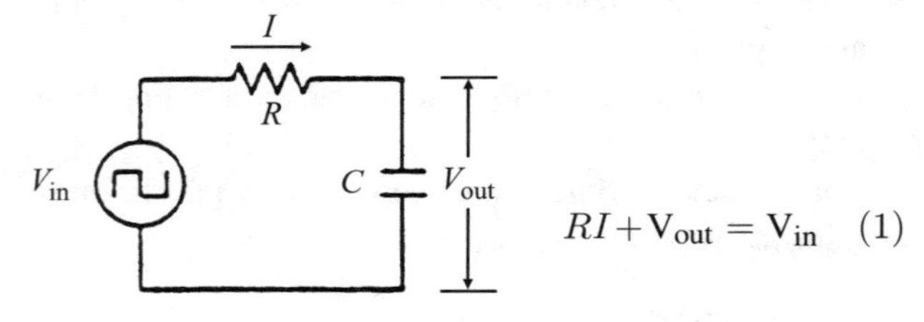

$$RI + \mathrm{V_{out}} = \mathrm{V_{in}} \qquad (1)$$

where R is the resistance and I is the current in the loop. Use the derivative form of Coulomb's Law to write $I = C\mathrm{V'_{out}}$, where C is the capacitance in the loop. Divide the terms of formula (1) by RC, and so we have the first-order linear ODE for $\mathrm{V_{out}}$:

$$\mathrm{V'_{out}} + \frac{1}{RC}\mathrm{V_{out}} = \frac{1}{RC}\mathrm{V_{in}} \qquad (2)$$

For simplicity, set $RC = 1$ and $\mathrm{V_{out}}(0) = 0$. Let's encode our message as a linear combination of square waves of various frequencies. The challenge is to find a range of frequencies for $\mathrm{V_{in}}(t)$ so that $\mathrm{V_{out}}(t)$ looks like $\mathrm{V_{in}}(t)$.

Figures 1 and 2 show the input voltage $\mathrm{V_{in}}(t)$ (dashed) and the output voltage $\mathrm{V_{out}}(t)$ (solid). We use $\mathrm{V_{in}}(t) =$ sqw $(t, 50, T)$, a square wave of period T and amplitude 1 which is "on" for the first 50% of each period. In Figure 1, T is 1 millisecond (so the frequency is 1000 hertz); in Figure 2, T is 25 milliseconds (40 hertz). The figures tell us we should encode our message with low-frequency voltages.

The classical solution formula for the linear ODE (2) wouldn't tell us much because of an awkward integration that involves the square wave, so we used a good numerical solver that likes square waves.

5. Oscillating Springs

Let's look at the way a spring-mass system responds to forces and initial conditions. Using Newton's Force Laws, we construct the modeling second-order ODE

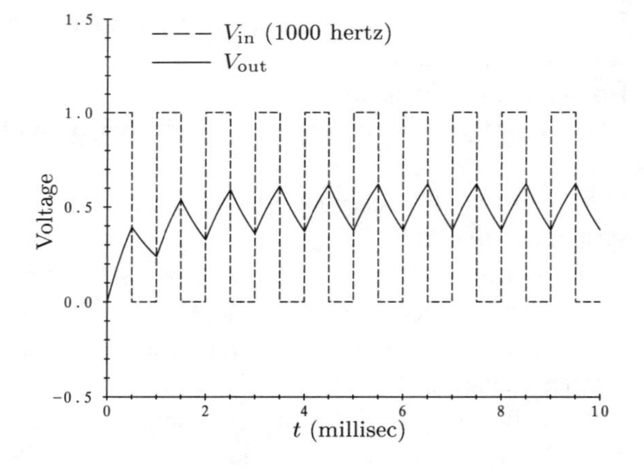

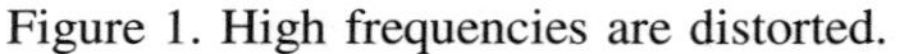
Figure 1. High frequencies are distorted.

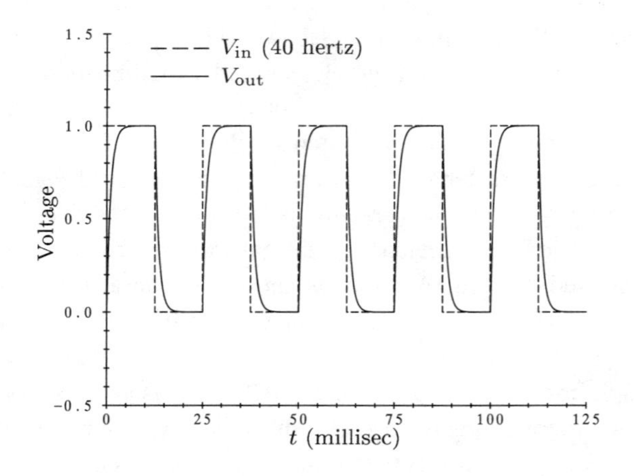

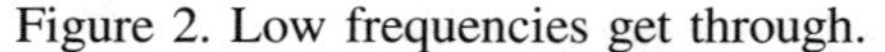
Figure 2. Low frequencies get through.

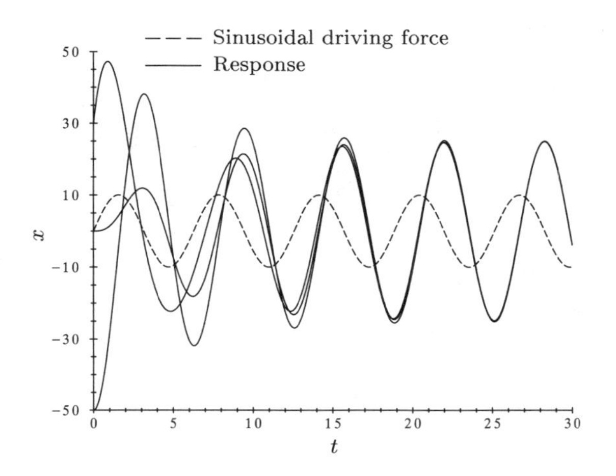

Figure 3. Periodic long-term behavior.

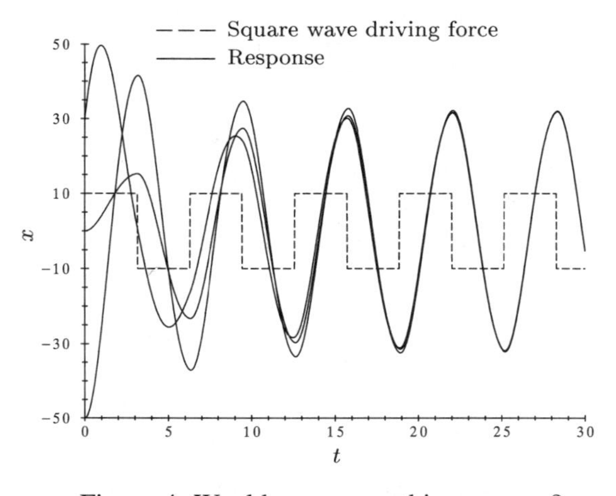

Figure 4. Would you guess this response?

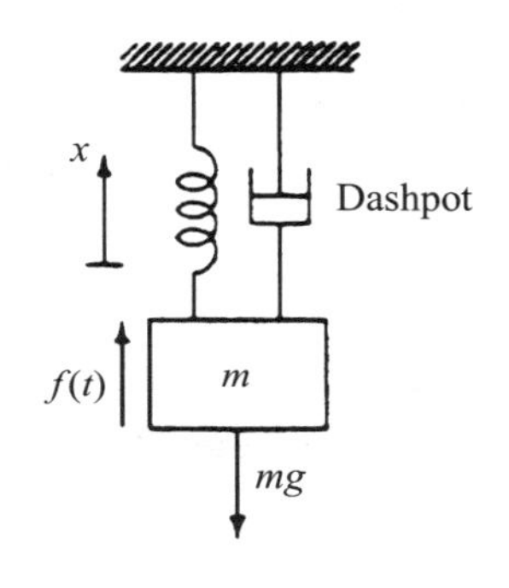

$$mx'' = S(x) - cx' - mg + f(t)$$
$$S(x) \text{ is the spring force} \quad (3)$$
$$c \text{ is the damping constant}$$

If $S(x)$ is given by $-ax$ (Hooke's Law), and x is re-measured from static equilibrium, the term $-mg$ is eliminated. Let's set $a/m = 1$, $c/m = 0.4$, $f(t) = 10 \sin t$ (Figure 3) or $f(t) = 20$ sqw $(t, 50, 2\pi) - 10$ (Figure 4), give $x(0)$ and $x'(0)$ various values, and use a numerical solver. All solutions are attracted with increasing time to a forced oscillation of the same period (2π) as the driving force. You would have trouble seeing this from a classical solution formula.

6. Soft Springs

Let's remove the driving force and suppose that the spring is "soft"; i.e., the spring force weakens with displacement from the neutral position. Measuring x from the end of an unstretched and uncompressed spring, we have the soft spring ODE,

$$mx'' = -ax + bx^3 - cx' - mg \quad (4)$$

where $-ax + bx^3$ models the nonlinear soft spring force. Figure 5 shows xx'-orbits if $a/m = 9.4$, $b/m = 0.2$, $c/m = 0.2$, and $g = 9.8$. The orbits near the right-hand equilibrium point (compressed soft spring) don't make any sense, but the oscillatory orbits near the middle equilibrium point (static equilibrium) are plausible. The orbits close to the left equilibrium point also have a story to tell. Is their soft-spring story fact or fiction? Figure 6 shows a solution curve of the nonlinear ODE (4) and a solution curve of the linearized ODE, $x'' = -9.4x - 0.2x' - 9.4$, both for $x(0) = -1$, $x'(0) = 5$, near the static equilibrium. Other than a slight phase shift, the linear ODE seems to be a good approximation to the nonlinear soft spring ODE near this equilibrium position.

Numerical solvers don't care whether ODEs are linear or nonlinear. But much more is known about linear ODEs than about nonlinear ones. It's reassuring to see that linear approximations can (sometimes) work quite well.

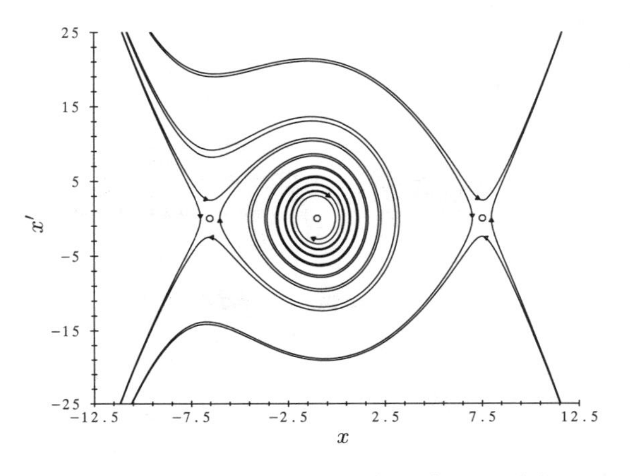

Figure 5. Orbits of a soft spring model.

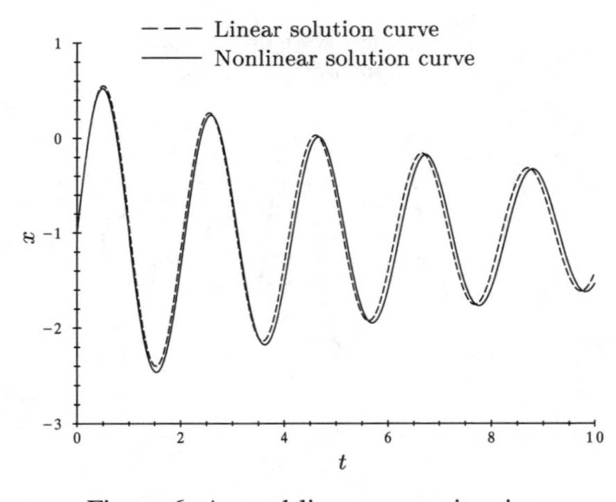

Figure 6. A good linear approximation.

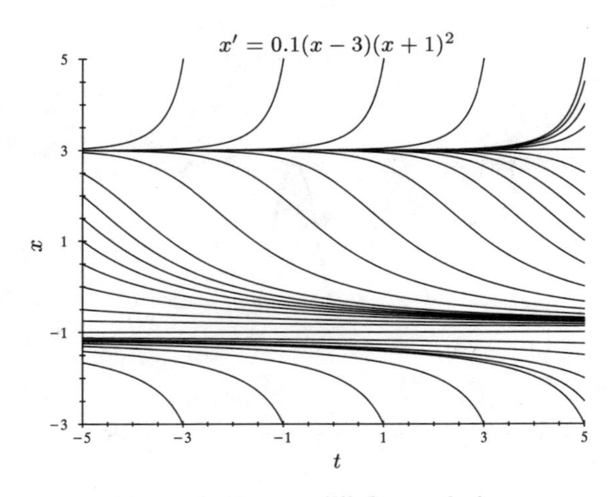

Figure 8. Two equilibrium solutions.

7. Sign Analysis

Autonomous first-order ODEs have the form $x' = f(x)$. A plot of the rate function $f(x)$ against x gives a great deal of information about the long-term behavior of the solution curves of the ODE. The zeros of $f(x)$ define equilibrium (constant) solutions, and the algebraic sign of $f(x)$ near one of its zeros tells us whether solution curves move toward or away from an equilibrium curve as $t \to +\infty$. Figures 7 and 8 show equilibrium solutions that attract or repel all nearby solution curves, or attract from one side and repel from the other.

8. Bifurcation, Harvesting, and Logistic Change

A logistic population $x(t)$ (think "fish") changes according to the nonlinear first-order ODE, $x' = r(1 - x/K)x$, where r and K are positive constants. The ODE, $x' = r(1 - x/K)x + H$, where H is a negative constant, models a population being harvested at a constant rate (think "fishermen"). Rescaling time and population gives the ODE

$$x' = (1 - x)x + c \tag{5}$$

with the single harvesting parameter c. As harvesting becomes more vigorous, attracting and repelling equilibrium population levels move toward each other (Figures 9 and 10), merge at a critical c-value, and vanish with devastating consequences. The range $-0.5 \leq x < 0$ shown in Figures 9 and 10 has no physical reality (ghosts of departed fish?). Let's track the changes by graphing the rate function $f = (1 - x)x + c$ for various values of c (Figure 11). Analysis of the sign of f tells us that for $c > -0.25$ there is always an attracting equilibrium level, but if $c < -0.25$ there is no equilibrium at all, and the population is doomed. So $c = -0.25$ is the bifurcation

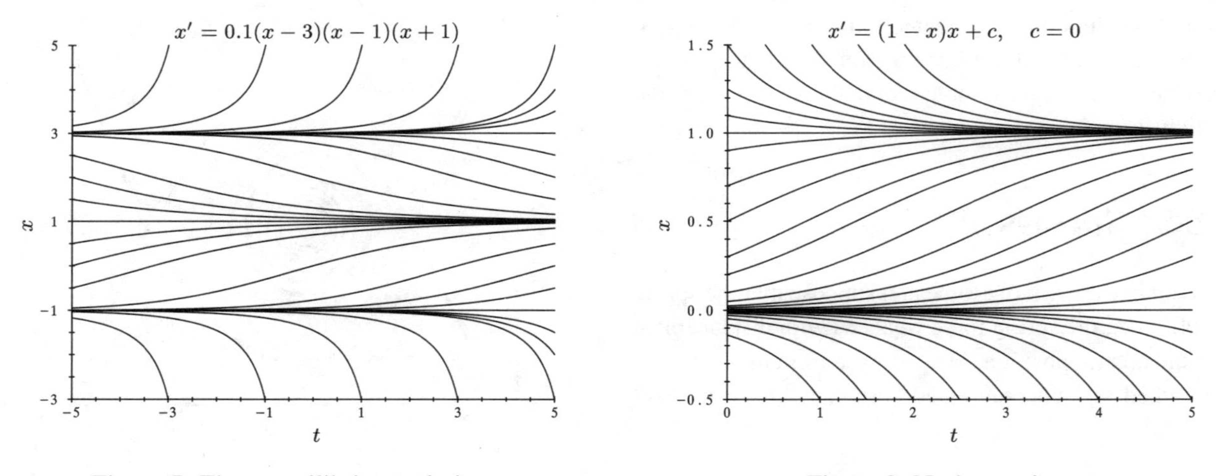

Figure 7. Three equilibrium solutions.

Figure 9. No harvesting.

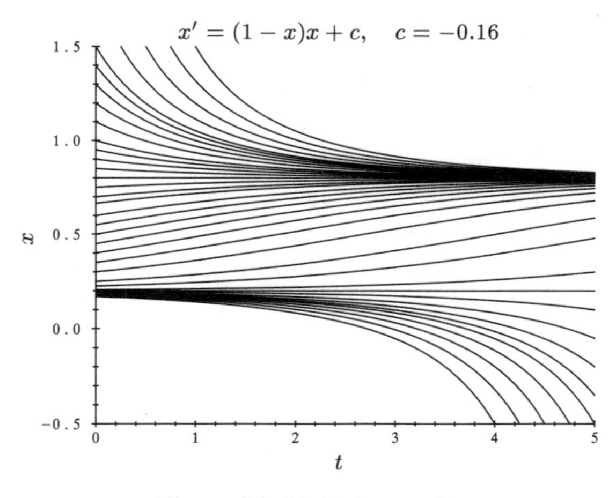

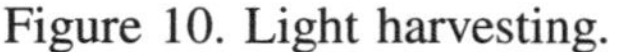

Figure 10. Light harvesting.

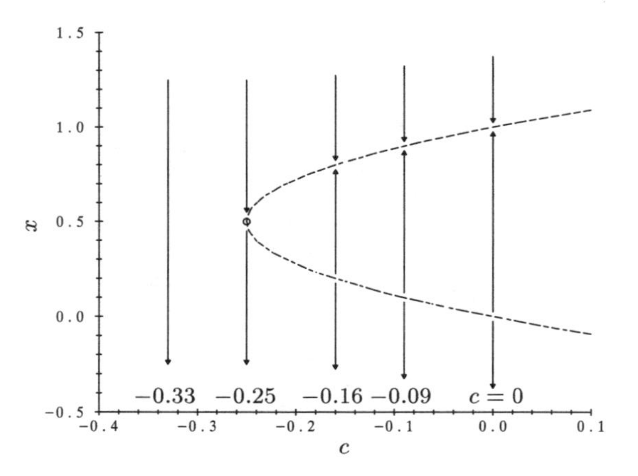

Figure 12. Bifurcation diagram.

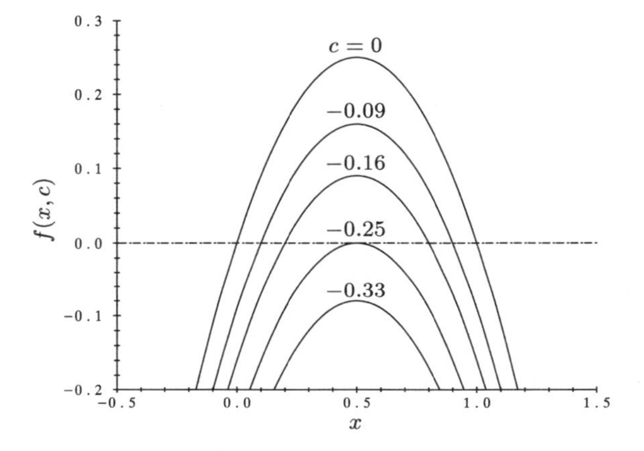

Figure 11. Preparation for sign analysis.

point between survival and extinction.

As c changes, equilibrium levels are tracked by the parabola, $(1-x)x + c = 0$ (Figure 12). The equilibrium points on the upper branch attract, those on the lower branch repel. The time-directed vertical "state-lines" are plots of $x(t)$ for various values of c. The bifurcation diagram of Figure 12 tells the whole story of bifurcation, harvesting, and logistic change: on the left we read disaster for the fish and, ultimately, for the fishermen, but in the middle we read good times for fish and fishermen. What's the fish story on the right?

9. Get the Lead Out

Lead enters our bodies via air, food, and water. The bloodstream distributes it to the bones and to the tissues where high lead levels are toxic. Lead leaks back into the blood from the bones (slowly) and from the tissues (rapidly), and is partially cleared from the blood by the kidneys. Hair, nails, and sweat also excrete small amounts of lead. The Environmental Protection Agency and the National Institutes of Health have done computer simulations and experiments with volunteers in Los Angeles and elsewhere to see just how lead flows through the body compartments, and how effective removal of lead from gasoline, paint, and other everyday items would be in lowering body lead levels.

We can model lead flow by the boxes and arrows of a compartment model. The boxes are the body compartments, the state variables x, y, and z are the amounts of lead in the compartments, and the labeled arrows give the entrance/exit rates of lead. The human studies showed that flow rate out of a body compartment is proportional to the amount of lead in the compartment. The balance law (rate of change = rate in − rate out) gives us the set of linear ODEs shown below.

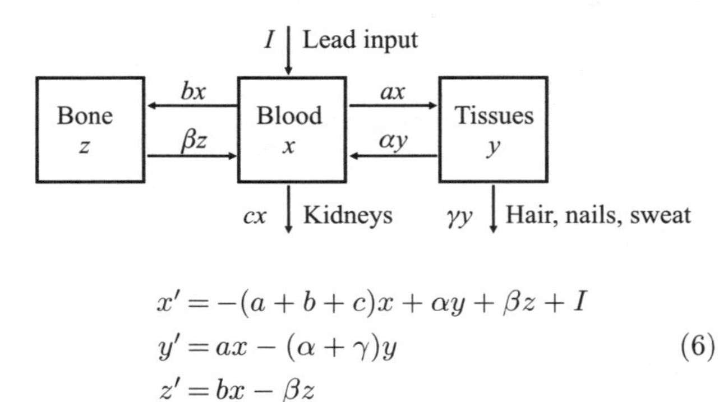

$$
\begin{aligned}
x' &= -(a+b+c)x + \alpha y + \beta z + I \\
y' &= ax - (\alpha + \gamma)y \\
z' &= bx - \beta z
\end{aligned}
\tag{6}
$$

Using the Los Angeles data, we have

$$a = 0.0111, \quad b = 0.0039, \quad c = 0.0211$$

$$\alpha = 0.0124 \quad \gamma = 0.0162, \quad \beta = 0.000035, \quad I = 49.3$$

where lead is measured in micrograms, time in days, the

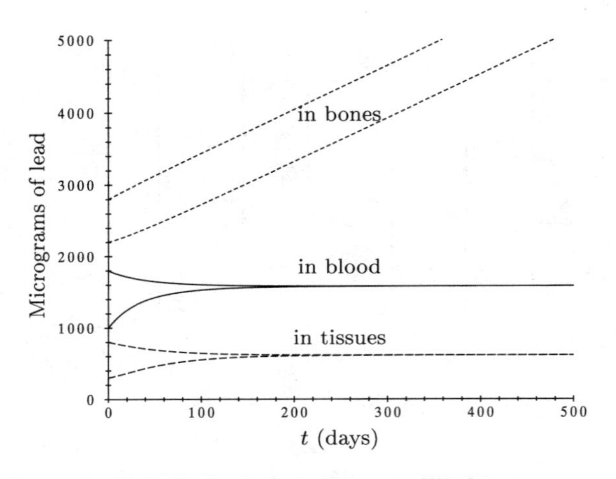

Figure 13. Approach to equilibrium.

units for a, ..., γ are $(\text{days})^{-1}$, and the units for I are micrograms/day.

The component curves in Figure 13 for two sets of initial data show that the lead in the blood and tissues soon approach equilibrium levels. The leakage rate from the bones back into the blood is so small that even after 500 days, the lead levels in the bones aren't even close to equilibrium (in fact, it would take more than 200 years to reach equilibrium). Figure 13 shows the high lead levels of the "bad old days" before lead was removed from gasoline, paint, and other products. Figure 14 shows what would happen if removal of lead from the environment reduced the lead ingestion rate I from 49.3 to 34.7 micrograms/day for all $t \geq 250$. And that is just what has happened in Southern California in the "good new days."

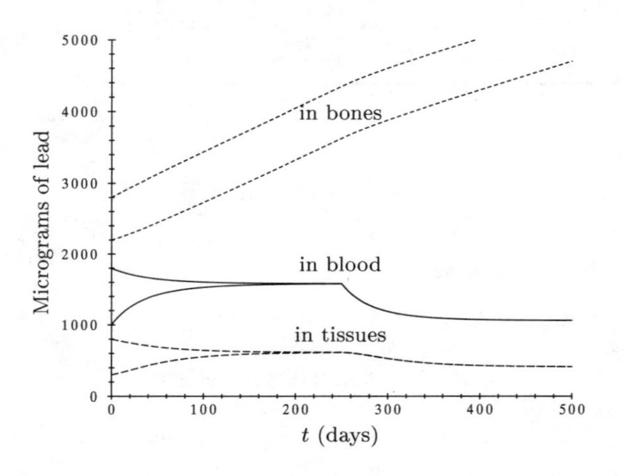

Figure 14. Lead-free gasoline, $t \geq 250$.

10. Strange Oscillations in a Chemical Reactor

Say that a reactant species in a chemical reaction changes into intermediate species, and the intermediates into a final product species. We can adapt the compartmental model idea and the balance law to model this process. In the reaction diagrammed below, reactant R becomes the intermediate A. As the reaction proceeds, A becomes B, and B turns into the final product P. Let's suppose (for now) that the reactions are first order, and so the transformation rate of a chemical is proportional to its concentration in the reactor. This gives us a set of four linear rate equations in the concentrations.

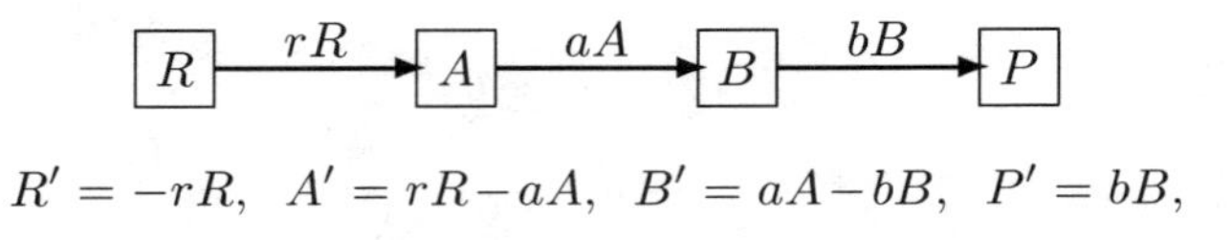

$$R' = -rR, \quad A' = rR - aA, \quad B' = aA - bB, \quad P' = bB,$$

where capital letters denote both a chemical species and its concentration, and lower case letters are rate constants. Concentrations and time are measured in dimensionless units for computational ease.

In recent times autocatalytic reactions have become commonplace. In autocatalysis a chemical species promotes its own production. Here's an example: two units of B react with one of A to produce three units of B, a net gain of one unit of B. By the Chemical Law of Mass Action the rate of decrease of A in this autocatalytic step is αAB^2 (α is a rate constant), while the rate of increase of B is $3\alpha AB^2 - 2\alpha AB^2 = \alpha AB^2$. Adding this autocatalytic step and starting with an initial concentration of R, we have a nonlinear initial value problem:

$$\boxed{R} \xrightarrow{rR} \boxed{A} \overset{aA}{\underset{\alpha AB^2}{\rightrightarrows}} \boxed{B} \xrightarrow{bB} \boxed{P}$$

$$\begin{aligned} R' &= -rR, & R(0) &= R_0 \\ A' &= rR - aA - \alpha AB^2, & A(0) &= 0 \\ B' &= aA + \alpha AB^2 - bB, & B(0) &= 0 \\ P' &= bB, & P(0) &= 0 \end{aligned} \qquad (7)$$

Figure 15 shows what happens if $r = 0.002$, $a = 0.08$, and α and b are scaled to 1.

Figure 15 shows unusual and unexpected oscillations in the concentrations in the intermediates if $R_0 = 540$. These oscillations are internally generated and are not due to oscillations in external factors such as temperature or pressure. Curiously enough, we can prevent the oscillations from turning on by heating the reaction pot, thereby increasing the coefficient a from 0.08 to 0.15 (Figure

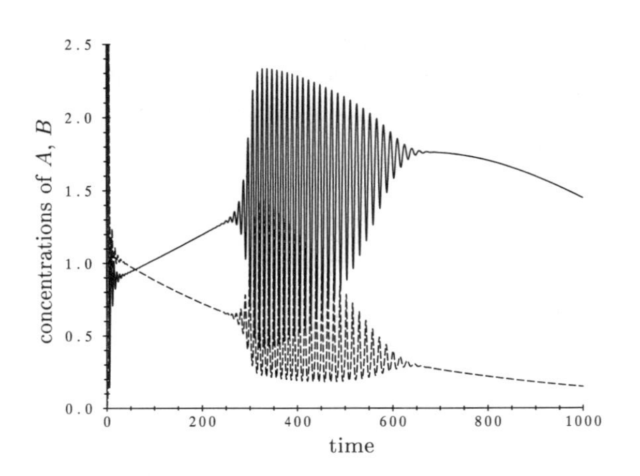

Figure 15. Autocatalytic reaction.

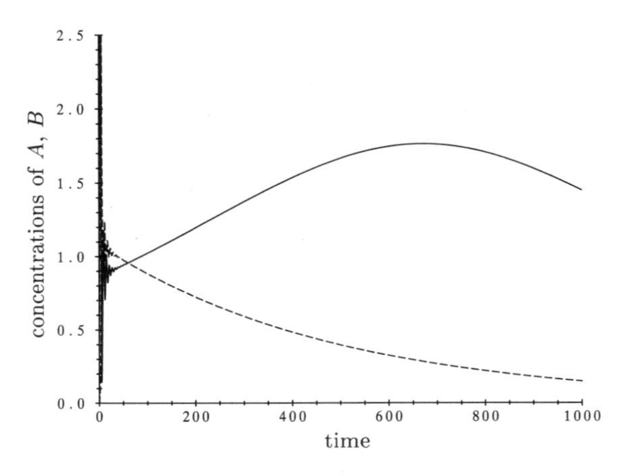

Figure 17. Numerics stops oscillations!

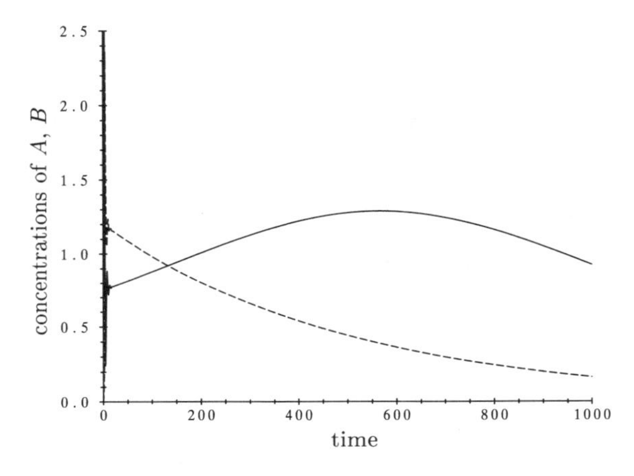

Figure 16. Heat stops oscillations.

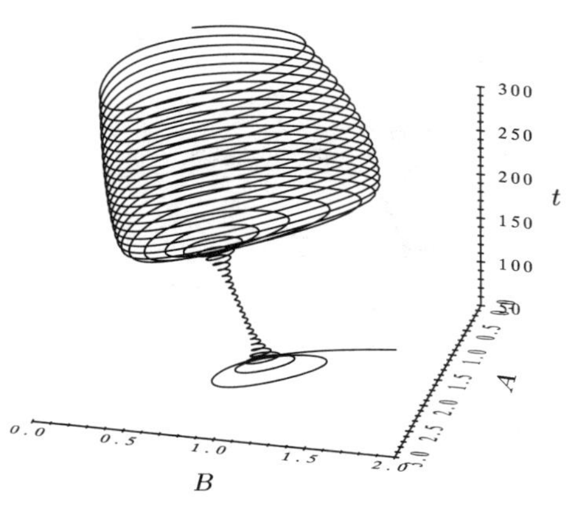

Figure 18. Intermediates time plot.

16). In a bizarre turn of events which should serve as a warning to be sceptical of what our computers tell us, we can also keep the oscillations from turning on by relaxing the numerical solver's internal error bounds (Figure 17). In fact, when we first simulated this system on our computers so that we could verify what chemists claimed, we didn't see any oscillations at all because we had set the solver's error bounds too high.

Figure 18 shows a plot of A and B against time if $R(0) = 500$. These oscillations are beginning to be understood chemically. Mathematically, it seems that the A, B oscillations of (7) are triggered by a bifurcation when R drops to around 370 and then turned off when R goes below 180. Put your solver to work on this system and see what happens if R_0 is 50, 100, 200, 500, 800—you are in for some surprises!

11. Stabilizing a Run-Away System

The orbits of the linear system, $x' = x + y$, $y' = -x$, spiral outward from an unstable equilibrium point and toward infinity (Figure 19). Throw in the single nonlinear term $-x^3$, and we transform the system into the stabilized van der Pol system

$$x' = x + y - x^3, \quad y' = -x$$

and create a periodic orbit that attracts all nonconstant orbits (Figure 20). This amazing feat is accomplished with forces that depend only on the state variable x and do not involve a time-periodic external driving force—small wonder that nonlinear models and controls are bread-and-butter tools of modern technology!

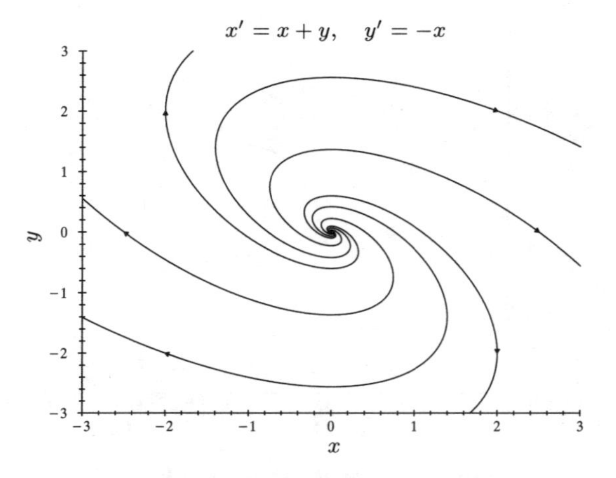

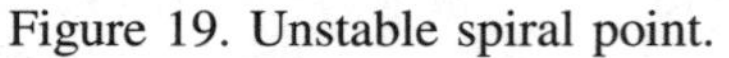

Figure 19. Unstable spiral point.

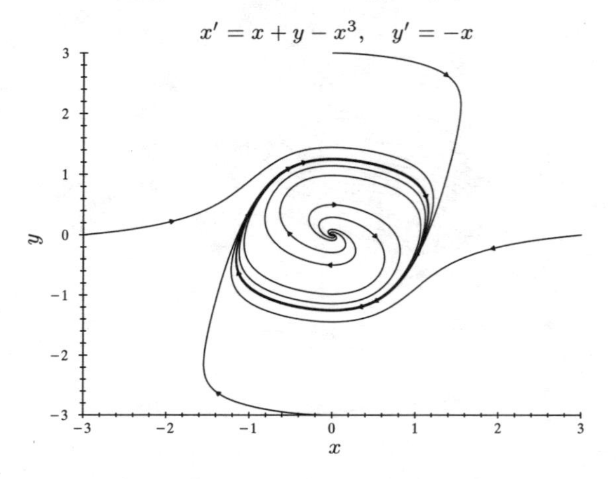

Figure 20. Attracting periodic orbit.

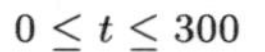

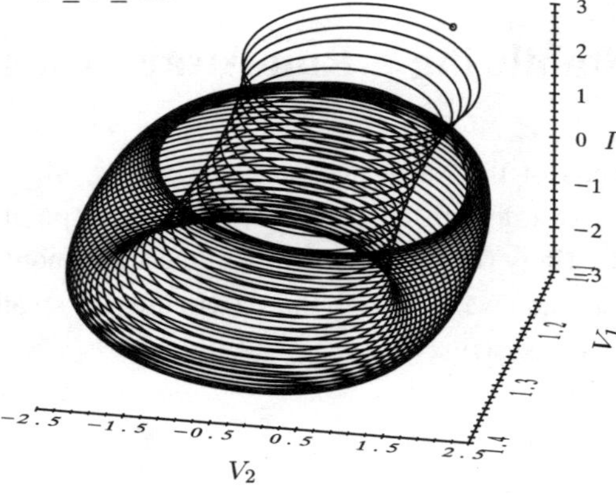

Figure 21. Scroll circuit orbit ($c = 0.1$).

12. How to Destroy a Stable Periodic Orbit and Create Chaos

Recently, electrical engineers have designed, built, and modeled a simple two-loop electrical circuit with a nonlinear resistor, two capacitors (one with a negative capacitance) and an inductor. Here is a nonlinear model initial value problem for the circuit that uses scaled current and voltages.

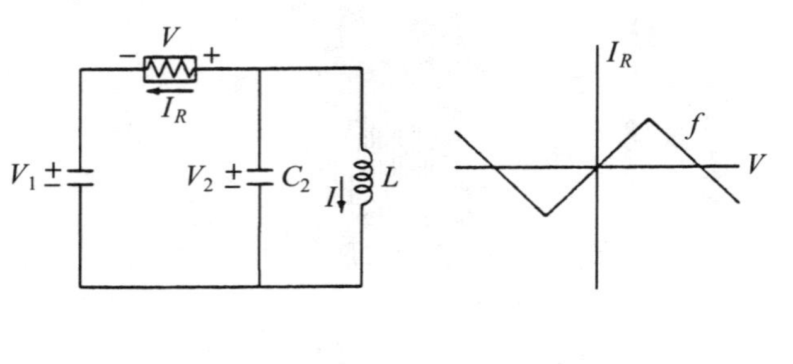

$$
\begin{aligned}
V_1' &= -cf(V_2 - V_1), & V_1(0) &= 1 \\
V_2' &= -f(V_2 - V_1) - I, & V_2(0) &= 1 \\
I' &= kV_2, & I(0) &= 1 \\
f(z) &= a - b(|z+1| - |z-1|)
\end{aligned}
\qquad (8)
$$

where $c = -C_2/C_1$, a, b, and k are positive circuit parameters. Experiments with the circuit suggest that $a = 0.07$, $b = 0.085$, $k = 1$ are realistic values. Let's see what happens to current and voltages when we change c.

At each low setting for c, there is an attracting periodic orbit (much like the van der Pol cycle) that attracts all nearby orbits as time advances. For $c = 0.1$ this cycle is just inside the lip of the mushroom orbit (Figure 21), an orbit which is attracted by the cycle. Project the mushroom into the V_1I-plane (Figure 22), and we see why the circuit is called a scroll circuit. The time span $0 \le t \le 300$ used for the orbit of Figure 21 was extended to $0 \le t \le 700$ to bring out the scrolling feature. Now "detune" the circuit by raising the value of c all the way up to 50, plotting orbits of (8) and their projections at each stage. You will see a strange and beautiful evolution away from the curve in Figure 21 and its projection

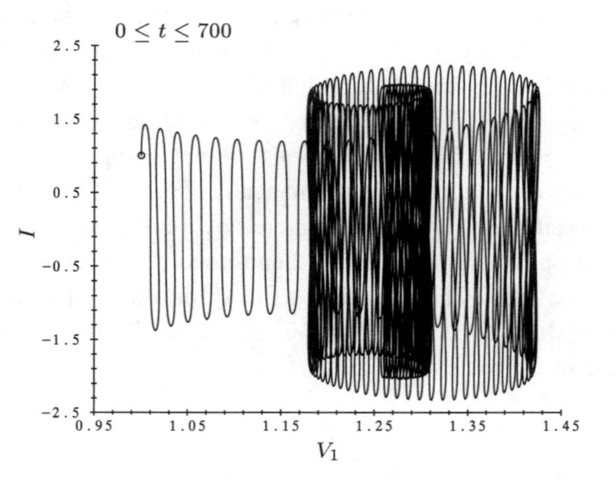

Figure 22. Projected orbit ($c = 0.1$).

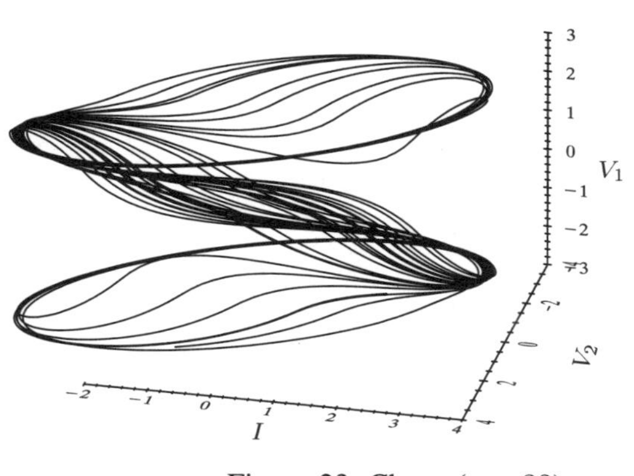

Figure 23. Chaos ($c = 33$).

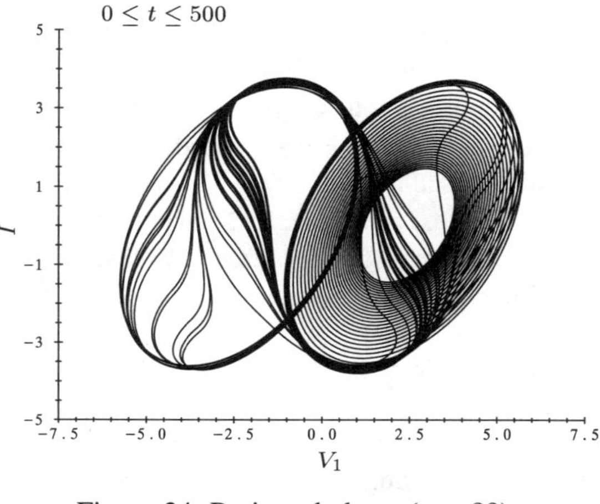

Figure 24. Projected chaos ($c = 33$).

in Figure 22. Around $c = 13$ the orbit begins to undergo seemingly random swings, and by $c = 33$ a completely different picture has emerged. Figure 23 shows the orbit of (8) for $c = 33$ plotted for $300 \leq t \leq 500$ and for a permutation of the axes of Figure 21 chosen to improve the view. This is a picture of "organized" chaos, the voltages and current of the circuit wandering at apparently random times across and between two distinct oval regions. Figure 24 shows the V_1I-projection over $0 \leq t \leq 500$.

Exploration of IVP (8) is a marvelous term computer project for a group of students (especially if they are potential engineers). The paper by Matsumoto listed in the Bibliography even gives simple instructions on how to build the circuit and display the voltages and current on an oscilloscope.

The scroll and related circuits are now at the heart of numerous engineering and mathematical studies aimed at understanding and explaining just what is going on. Mathematically, it appears that the increasing value of c triggers a sequence of bifurcations of various kinds that culminates in the chaotic but disciplined behavior captured by the numerical solver in Figures 23 and 24.

Conclusion

Until very recently, introductory courses in ODEs relied mostly on four approaches: clever tricks for finding solution formulas, some mathematical theory, some discussion of numerical algorithms for generating approximate solutions, and a treatment of several applications (mostly from physics). With the widespread availability of good numerical ODE solvers and inexpensive platforms the options for the course syllabus have multiplied. Computer simulations give us a superb tool for examining dynamical systems in ways we could previously only dream about. In addition, modeling is now a hands-on process for students because it is easy to explore different modeling elements and quickly see the results.

Computer images give the student a deeper understanding of the theory of dynamical systems and their properties. In the new approach theory, solution formulas, modeling, and computer simulations complement one another in a highly supportive way. We look forward to seeing the approach sharpened and developed into the next century.

Bibliography

General

1. R.L. Borrelli and C.S. Coleman, *Differential Equations: A Modeling Perspective*, John Wiley & Sons, Inc., New York, 1998.
2. ——, W. Boyce. *Differential Equations Laboratory Workbook*, John Wiley & Sons, Inc., New York, 1992.
3. J.H. Hubbard and B.H. West, *Differential Equations: A Dynamical Systems Approach*, Springer-Verlag, New York, 1991 (v. 1), 1995 (v. 2).

Newsletter

C·ODE·E is available on the Web: http:/www.math.hmc.edu/codee, but is no longer available in paper. The newsletter has articles on models and computer experiments with dynamical systems. Efforts to revive the letter are underway.

Sources

1. Batschelet, Brand, Steiner, *J. Math. Biol.*, **8** (1979), pp. 15–23 (for lead in the human body).

2. A. Felzer, "First Order Analysis of Communication Channels," C·ODE·E, Winter Issue, 1993.
3. P. Gray and S. Scott, *Chemical Oscillations and Instabilities*, Clarendon Press, Oxford, 1994 (for the autocatalytic oscillations).
4. T. Matsumoto, L.O. Chua, R. Tokunaga, *IEEE Transactions on Circuits and Systems*, CAS–34(3): 240–253, March 1987 (basic source on scroll circuits and how to build one).
5. T.S. Parker and L.O. Chua, *Practical Numerical Algorithms for Chaotic Systems*, Springer-Verlag, New York, 1989 (discusses some scroll circuits).
6. B. West, S. Strogatz, J. McDill, J. Cantwell, H. Hohn, *Interactive Differential Equations* (CD-ROM and Workbook), Addison-Wesley, 1996, 1997.

$mx'' = S(x) - cx' - mg + f(t)$
$S(x)$ is the spring force
c is the damping constant

Differential Equations in the Information Age

William E. Boyce
Rensselaer Polytechnic Institute

Abstract

The ready accessibility of graphing calculators, symbolic manipulators, and the World Wide Web has made possible very substantial changes in the traditional course on differential equations. These possible changes affect both the content and style of instruction in the course. This paper discusses some options that are now practical in both areas.

A differential equations course is an important element in the education of engineers, scientists, and mathematicians because of the manifold uses of the subject in studying the natural world. It is also an unsurpassed vehicle for demonstrating and learning the advantages and difficulties in formulating a problem in mathematical terms, carrying out a mathematical investigation of the mathematical model, and then drawing conclusions from the results.

Rapid and accelerating changes in the ability to transmit and to process information have resulted in an environment for teaching differential equations (or other courses, for that matter) that is fundamentally different from the one that existed only a few years ago. This environment continues to evolve at an extremely rapid pace.

There are at least three levels of change, corresponding to levels of technological usage:

1. graphing calculators — perform moderately extensive numerical calculations and plot relatively simple graphs. They are inexpensive and extremely portable.
2. symbolic manipulators — execute complex symbolic manipulations, as well as extensive numerical calculations, and produce outstanding graphical displays.
3. World Wide Web — provides easy access to materials developed by others at sites around the world.

These technological advances enable instructors, if they wish, to modify very substantially the traditional course in differential equations. These modifications relate both to the course content and to the manner in which the course is conducted.

Graphing calculators enable you to ask more quantitative questions, to consider problems closer to the real world, and to see the results graphically.

Symbolic manipulators greatly enhance these capabilities and, in addition, enable you to execute laborious and complicated symbolic algorithms that are needed to obtain either exact solutions or approximations.

The World Wide Web has the potential, already beginning to be realized, of enormously enlarging the range of materials that are easily available to enrich or to add variety to a course.

Effects on the content

In just a few words, the effect of technology is to make the course much more visual, much more quantitative, and much less formula-centered than the traditional course has been during the past several decades [1],[2]. Rather than focusing almost entirely on pencil and paper methods for solving those equations that can be solved in this manner, the course can dwell less on such methods and can be expanded to include other elements. Many relatively simple-looking problems, of course, have solutions that cannot be expressed in terms of elementary functions and for such problems, numerical or graphical methods are needed. Moreover, even if a problem can be solved in terms of elementary functions, numerical computations or plots may be required in order to interpret the solution or to draw interesting conclusions from it. The important role of parameters in a problem becomes much more accessible. In fact, it is often more important and interesting to understand how a solution depends on a parameter than to find the solution for some particular value of the parameter. Finally, a course on differential equations affords many opportunities to practice mathematical modeling, an area in which most students need all the practice they can get.

The following examples illustrate some of these observations.

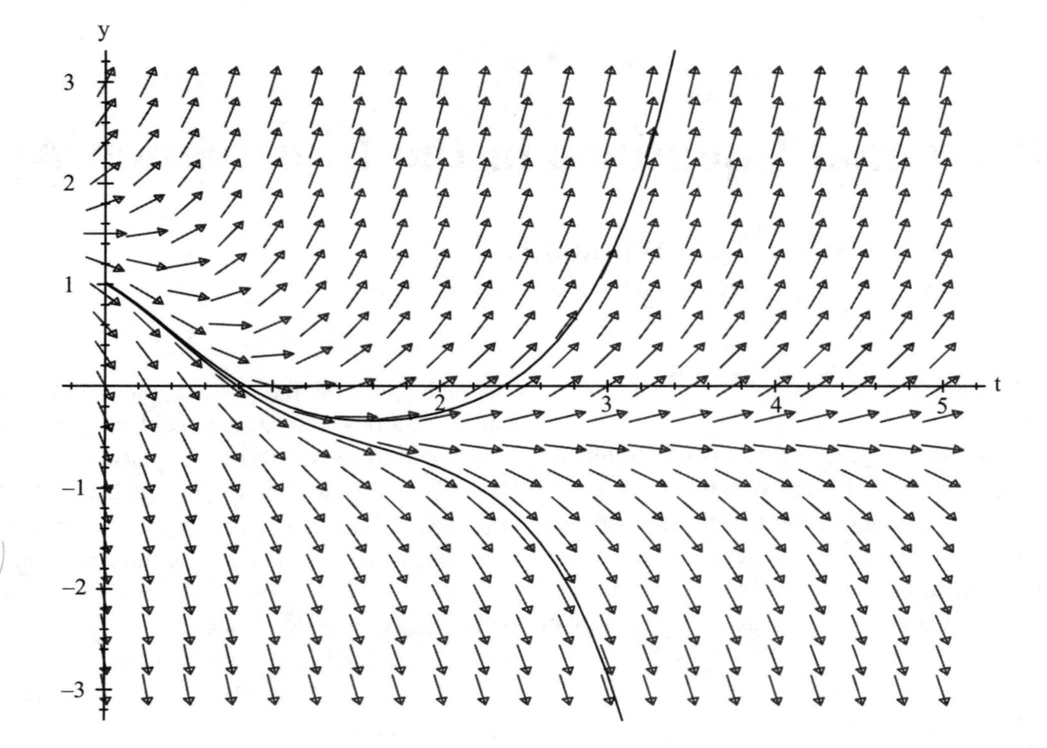

Figure 1:

Example 1. Find the solution of

$$y' - 2y = 1 - 4e^{-t}, \qquad y(0) = 0, \tag{1}$$

and describe its behavior for large t.

The general solution of the differential equation is

$$y = ce^{2t} - \frac{1}{2} + \frac{4}{3}e^{-t}. \tag{2}$$

The initial condition is satisfied only if $c = -5/6$, so in this case the solution becomes negatively unbounded as $t \to \infty$. By examining the general solution, or by using other initial conditions, you can see that sometimes the solution grows negatively (if $c < 0$) and sometimes it grows positively (if $c > 0$). The transition from one type of behavior to the other occurs when $c = 0$, that is, when $y(0) = 5/6$. Thus $5/6$ is the *critical initial value* that separates solutions that behave in two quite different ways. We use problems similar to this one early in a differential equations course not only to show how to solve first order linear equations, but, of more importance, to begin to accustom students to the idea that they should always seek to interpret the solutions that they obtain. In particular, it is often important to determine the long-time behavior of the solution, and to relate it to the initial state.

If you wish to make the problem slightly more complicated, you can replace the given equation by

$$y' - 2y = 1 - 4e^{-t^2}. \tag{3}$$

In this case the solution involves the error function, so an analytical discussion may be out of place in an elementary course. Nevertheless, a direction field for the equation can be readily constructed (see Figure 1), and it strongly suggests that again there is a critical initial value that separates solutions that grow positively from those that grow negatively. One can estimate this critical initial value from the direction field, or more accurately, by plotting a few numerically generated approximations to solutions. For example, Figure 1 shows the solutions that satisfy the initial conditions $y(0) = 1.01$ and $y(0) = 1.02$, respectively. Based on this figure, it seems clear that the critical initial value is between these two values. A bisection method can be employed to estimate the critical initial value as accurately as needed. Alternatively, one can use a symbolic differential equation solver to find the symbolic solution, and from it determine the critical initial value either exactly, or to as many decimal places as needed.

Example 2. The initial value problem

$$u'' + 0.2u' + 2u = 3\cos t, \quad u(0) = 0, \; u'(0) = 1, \tag{4}$$

is typical of those that describe a periodically forced linear oscillator, such as a spring-mass system or a simple pendulum. By using straightforward methods covered in almost every course on elementary differential equations one can obtain the solution

$$\begin{aligned} u &= -\frac{75}{26}e^{-t/10}\cos\frac{\sqrt{199}}{10}t \\ &\quad +\frac{35\sqrt{199}}{5174}e^{-t/10}\sin\frac{\sqrt{199}}{10}t \\ &\quad +\frac{15}{26}\sin t+\frac{75}{26}\cos t. \end{aligned} \tag{5}$$

However, the algebraic manipulations required to reach this solution are complicated enough to cause significant difficulties for most undergraduate students of engineering or science. They are certainly complicated enough and time-consuming enough to make it unlikely that a student would solve very many problems of this kind by hand. Many students would never be able to get to the solution and be confident that it was correct.

Yet what one would like to do is to draw some conclusions about the underlying physical problem that gave rise to the initial value problem (4). For example, the first two terms in the solution (5) constitute the transient part of the solution. How long does it take for the transient to become insignificant, less in magnitude than some given ϵ, perhaps? Or, what is the amplitude and phase of the steady state part of the solution, and how do they compare with the amplitude and phase of the forcing term? Such questions become much more accessible if one's time and energy are not wholly consumed just in obtaining an expression for the solution.

In fact, technology now enables you to go much further. For instance, you can replace the forcing term in (4) by $\cos\omega t$ and obtain the solution in terms of ω. The preceding questions now can be modified to inquire how the decay time of the transient and the amplitude A of the steady state response depend on ω. If you ask for a plot of A versus ω, students will see that there is a peak response A^* occurring at a resonant frequency ω^*, which is close to the natural frequency ω_0. If you wish, you can go one step further yet, consider the mass, drag coefficient, or spring constant as parameters, and ask how A^* depends on these quantities.

While all of these things can be done, and have been done, by hand, today's technology makes them accessible to average engineering or science students as a routine part of an elementary differential equations course. At Rensselaer we regularly assign problems such as the one described in this example as homework problems in our course. After a while students get the idea that they can actually use mathematics to solve problems related to their own interests in engineering or science.

Example 3. The Lotka-Volterra, or predator-prey, equations are often used to illustrate nonlinear autonomous systems. A typical system is

$$dx/dt = x(2-y), \quad dy/dt = y(-3+x), \tag{6}$$

where x and y are the populations of the prey and predator, respectively.

It is well-known that trajectories in the first quadrant of the xy plane are closed curves surrounding the stable equilibrium point, which is $(3,2)$ in (6). These trajectories correspond to periodic variations in the populations x and y with time. Students can readily verify these facts by plotting one or more trajectories in the xy plane, along with the corresponding plots of x versus t and y versus t. Figures 2, 3, and 4 show these plots for the initial conditions $x(0) = 5$, $y(0) = 3$. From Figures 3 and 4 students can read off estimates of the period and amplitudes of the population variations. By choosing different initial conditions they can investigate how the period relates to the amplitude for the prey and predator.

At the next level, one can replace one or more of the numerical coefficients in (6) by a parameter, for instance,

$$dx/dt = x(2-\alpha y), \quad dy/dt = y(-3+x), \tag{7}$$

where the positive parameter α is the predation coefficient. Then the question is to investigate how the period and amplitudes depend on α.

Finally, students sometimes have trouble relating the time responses (Figures 3 and 4) with the phase plot (Figure 2). To make the relation clear it is helpful to ask them

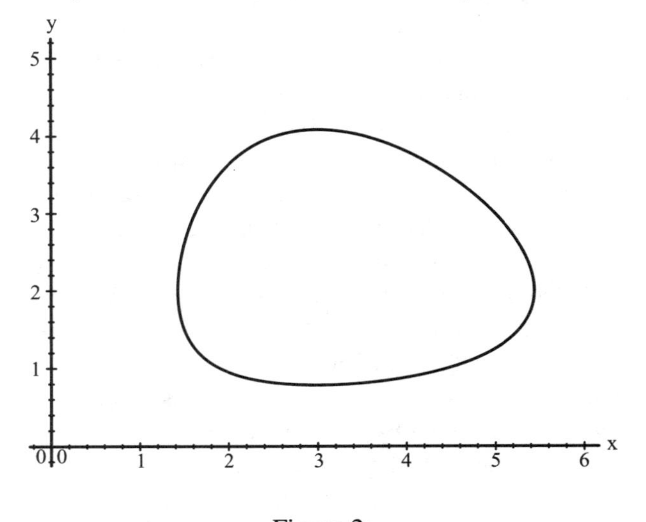

Figure 2:

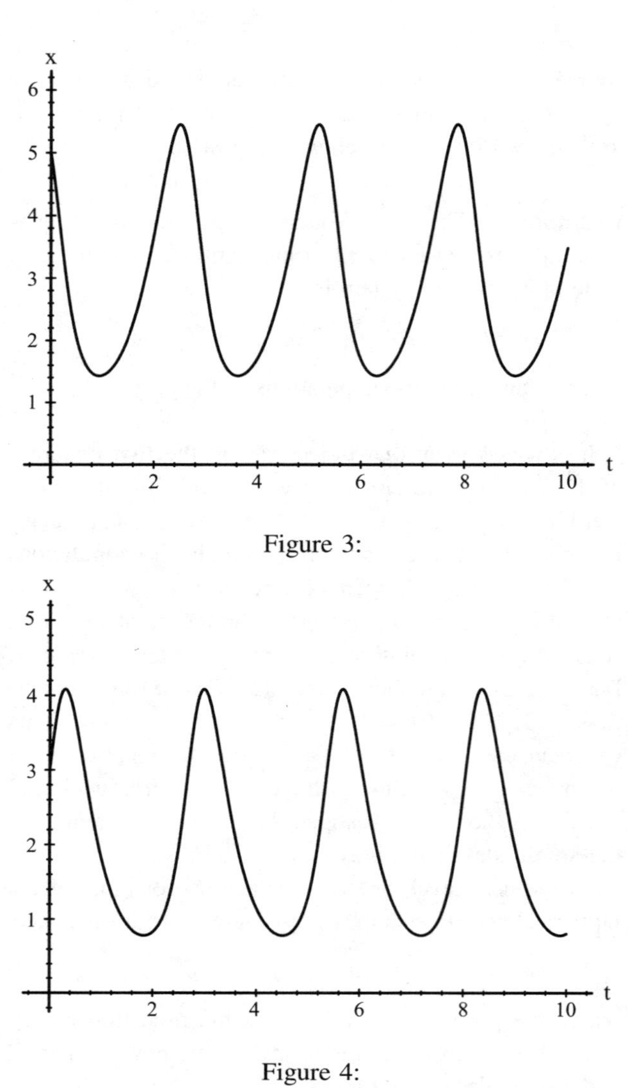

Figure 3:

Figure 4:

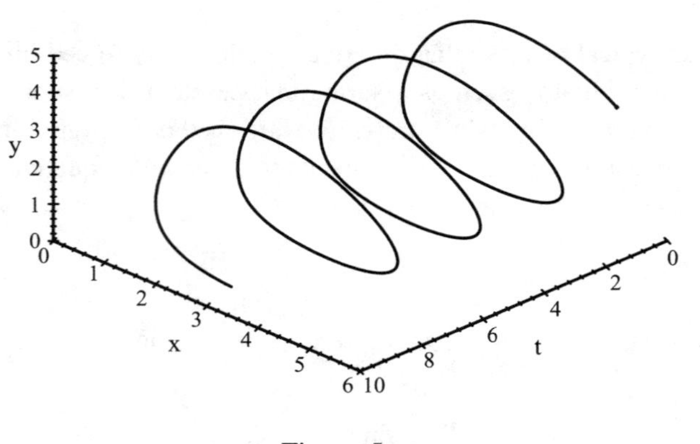

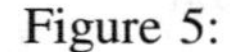

Figure 5:

to plot the solution in three-dimensional txy space (Figure 5). By suitably rotating this figure, one can in effect project the three-dimensional curve into each of the coordinate planes, thereby obtaining Figures 2, 3, and 4 once again. This clinches the fact that these figures are just different views of the same solution.

As these examples show, you (the instructor) are no longer limited to model problems that are easily solvable by hand by inexperienced students. With readily available computer assistance students can draw direction fields, compute and plot solutions of more complicated equations, and examine models that are closer to the real world and therefore have greater intrinsic interest.

As an instructor, I believe that we should take advantage of this situation by spending less time on paper and pencil solution methods. The computer can perform these methods faster and more accurately, and in general we should not be teaching students to do what a computer can do. The computer will win every time. Instead, we should be teaching them to do things that computers cannot do now, and probably not in the near future: such things as mathematical modeling, the interpretation of results, and the design of systems that will behave in specified ways. With respect to the time-honored classical solution methods the real question is—how much emphasis on these methods is needed in order to learn what differential equations are all about? To put the question in more concrete terms—how many times does a student need to find an integrating factor in order to understand how to solve first order linear equations? How many times does a student need to factor a characteristic polynomial in order to understand the structure of solutions of second order linear homogeneous equations with constant coefficients? How many times does a student need to evaluate arbitrary constants from initial conditions to understand this process?

We want students to gain a correct conceptual understanding of how to solve the important problem types, and this is achieved, or at least reinforced, by carrying out the details in a few problems by hand. But once the understanding is achieved, we should let a computer do the detailed calculations and turn our attention to other matters, such as whether our model is an adequate reflection of physical reality, and whether we can draw any useful conclusions from the results of our analysis.

Effects on the style of instruction

An important aspect of a course on differential equations is the transmission of information, a body of knowledge that has promise of being useful in the students' careers, or perhaps just interesting to explore. In the past, the main sources of the information were the instructor and the textbook. Today, with easy access to the World Wide

Web and to a growing number of skillfully produced CDs, both students and instructors have immediate access to an enormous quantity of information about differential equations.

For instructors this means that they have the opportunity, if they choose, to incorporate some of this material into their course. Naturally, there may be some difficulties involved in doing this, perhaps some discontinuities of approach that need to be smoothed over, and each instructor must weigh the potential benefits against the costs.

For students the possibility exists of finding something exciting that they would not have encountered otherwise in their course. However, the lack of structure and order on the web often makes it difficult for students to evaluate and appreciate the information they may encounter.

For both students and instructors the sheer volume of available information raises the questions of how to find efficiently the specific facts or methods that are needed for a given problem and how to organize the selected material coherently.

As it becomes possible to use technology to perform some of the instructional tasks that can be automated, including some of the information transmission, the role of the instructor needs to evolve. Freed to some extent of the responsibility of being the primary source of information, the instructor can spend more time on other activities, such as contact with individual students or small groups. The instructor can become a guide who helps students to learn how to navigate on the sea of electronic information with some discrimination and confidence. The instructor also needs to provide a structure relating various topics in the course to each other and to other areas of study. Of course, the instructor must continue to help to motivate students to put forth a serious effort in the course, and generally to provide guidance and encouragement to students struggling to master unfamiliar and often challenging material—to fulfill a cheerleading function, so to speak.

Finally, the course itself needs a transformation—away from more or less comprehensive coverage of a certain restricted set of topics, illustrated by carefully chosen examples whose solutions do not exceed a fairly low threshold of difficulty, and toward the investigation of more extensive and partially open-ended problems. The focus shifts somewhat away from a thorough treatment of certain algorithmic methods and toward the process of mathematical investigation of natural phenomena. This process involves the formulation of a problem in mathematical terms, its mathematical solution, interpretation of the results, comparison with reality, reformulation of the problem, more analysis, and so forth, until the results are satisfactory. This process becomes part of a student's mindset and can be used to advantage in problems far from those on which it was learned. On the other hand, details of algorithmic methods are quickly forgotten unless used often. This is not necessarily critical because such details can be looked up again if needed, or (now) are often incorporated into computer software.

Traditionally, courses in differential equations have been taught in a lecture mode. Where classes are small, instructors can become less formal, and can often stimulate class discussion that is lively and productive. However, in many institutions the class size has slowly crept upward to the point that real discussion is difficult and more formal lectures have dominated, sometimes accompanied by weekly meetings for recitation or problem discussion in smaller groups. In recent years, many colleges have introduced laboratory meetings, in which computer explorations can be carried out, along with the traditional lectures and recitations. A few have begun to try out more novel arrangements. For example, at Rensselaer we have begun to offer courses in several different departments in what we call a studio format, in which the distinctions among lecture, recitation, and laboratory are reduced or obliterated.

Studio classes vary somewhat from subject to subject and from instructor to instructor ([3],[4]) but typically they rest on two principal ideas. One idea underlying a studio type course is to combine and integrate the various elements of the course (lecture, recitation, discussion, laboratory) into a single entity, so that they will reinforce each other, rather than appearing to the student as disparate and sometimes almost unrelated activities. This requires a reconfiguration of space into rooms that are suitably equipped for a variety of activities. In differential equations this means that class meetings should be held in a room that permits each student to have access to a computer at all times. At Rensselaer we are gradually moving away from a model based on classrooms outfitted with fixed workstations to a model based on classrooms equipped with a power source and an Ethernet connection for each student. Students bring their own laptop or notebook computers to class, plug them in, and are ready to go to work.

The other main idea underlying a studio course is that more learning takes place when students are thinking about and doing mathematics, rather than when they are watching and listening while the instructor does mathematics. In other words, the focus is on the students and their activities, rather than on what the instructor is doing. Consequently, a studio classroom is a rather unstructured place. Groups of students may be working on a problem, some may be using computers, others may be talking with each other or with the instructor. All this takes place interspersed with short presentations by the instructor or by students, punctuated by discussion involving the whole class.

For an experienced instructor it is relatively easy to lecture for fifty minutes on some familiar topic in differential equations. The decision to teach a course in a studio style requires some substantial changes, and is a decision not to be taken lightly. In the first place, it is not easy to design activities that have a chance of capturing the interest of most students, that will keep them busy most of the time, that allow for different levels of proficiency, and that enable the class to cover the subject matter in the time allotted to the course. Further, in a studio class the instructor is no longer in complete control of what is going to happen, but must be flexible enough to orchestrate a variety of activities and to seize opportunities as they occur, sometimes unexpectedly. However, the good news is that you have a lot more close contact with individual students and this makes a studio course very satisfying for an instructor.

In addition to appropriate space and a willing instructor, a successful studio class requires instructional materials designed for this type of course. Commercial publishers (for example, [5],[6]) are beginning to provide materials that fit very nicely with the studio concept. The Web [7] is another rich source of materials, often placed there by instructors who have developed the materials originally for their own use. The National Science Foundation, through its Mathematics Across the Curriculum program, is underwriting the creation of a wide variety of materials. Under this NSF program there is a project based at Rensselaer, Mathematics and its Applications in Engineering and Science: Building the Links (known informally as Project Links), that is generating a library of interactive multimedia modules linking topics in mathematics with current applications in engineering and science. One of the principal areas of interest is differential equations. As it is developed, this material will be available on the Web [8].

As technological advances spread ever more widely through the educational system, and as instructional materials designed to take advantage of them become more numerous, instructors have many opportunities to make the study of differential equations more up-to-date, more interesting, and more useful to their students. The opportunities and the corresponding challenges they present make teaching a course in differential equations more exciting now than ever.

References

1. The College Mathematics Journal, Special Issue on Differential Equations, Vol. 25, No. 5, November 1994. The articles by William Boyce, Paul Blanchard, and Paul Davis are particularly relevant to the question of overall course content.
2. Robert L. Devaney, *The Future ODE Course is Already Here*, Consortium for ODE Experiments (C∗ODE∗E) Newsletter, Winter, 1996, pp. 2–3.
3. Jack M. Wilson, The CUPLE Physics Studio, *The Physics Teacher*, Vol. 32, December 1994, pp. 518–523.
4. Joseph G. Ecker, Preface to *Studio Calculus*, HarperCollins College Publishers, New York, 1996.
5. Beverly West, Steven Strogatz, Jean Marie McDill, and John Cantwell, *Interactive Differential Equations*, Addison-Wesley Interactive, 1996.
6. Consortium for ODE Experiments (C∗ODE∗E), ODE Architect, John Wiley and Sons, Inc., New York **(to appear)**.
7. Individual sites are too numerous to mention here. A good source of information is the Consortium for ODE Experiments (C∗ODE∗E), based at Harvey Mudd College. Its URL is `http://www.math.hmc.edu/codee/`.
8. The URL for Project Links is `http://links.math.rpi.edu`.

A Geometric Approach to Ordinary Differential Equations

Michael Branton and Margie Hale
Stetson University

1. Introduction

Recent changes in the content and presentation of the first course in Ordinary Differential Equations are moving it from a course in which a great deal of content was focused on methods of solution, to a course that has a greater focus on employing geometric ideas to help in the understanding and analysis of the equations and their solutions. This is evidenced by recent texts such as those of Hubbard and West [HW], Borrelli and Coleman [BC], or Blanchard, Devaney, and Hall [BDH].

There are at least three influences responsible for an increased emphasis on a geometric point of view: The integration of a Dynamical Systems perspective and its emphasis on qualitative analysis, the influence of the Calculus reform movement and its Numerical, Analytical and Graphical philosophy, and the increase in easily available computer technology which enables almost any student to view and manipulate geometric representations of differential equations.

The geometric viewpoint will be evidenced in this paper in three main ways. First, we will interpret a differential equation or a system of differential equations as containing geometric information in the form of a slope field or vector field, and use this as a primary means of understanding the equation(s). Second, we will be interested in a qualitative analysis of solutions, rather than in calculating analytic solutions. Third, we will, whenever possible, want to emphasize visualization. None of this is meant to say that an analytic approach is unimportant, however, it is not our business in this paper.

We hope to guide the reader through a number of the insights and delights which this approach can bring. In some of what follows, we will presume as a minimal background for a student a course in multivariable calculus and an introductory course in Linear Algebra. We include many definitions and some references to aid in further reading.

2. First-order Equations

2.1 The Slope Field

A first-order ordinary differential equation defines, either explicity or implicitly, a differentiable function

$$F : D \to \mathbb{R},$$

where D is an open subset of $\mathbb{R}^2$. For example, in the linear equation

$$t\frac{dy}{dt} + 2y = t^3,$$

the function F is found by solving for the derivative:

$$F(t, y) = \frac{dy}{dt} = \frac{-2y}{t} + t^2.$$

The function F is defined on $\{(t, y) \in \mathbb{R}^2 : t \neq 0\}$.

The **slope field** of the differential equation is a graph of the function F in the following sense. For each point (t, y) of the domain, F defines the slope of any solution y which passes through the point (t, y). That slope is represented by a small line segment having the requisite slope and drawn with its center at (t, y). See Figure 2.1.1 below. (Most modern differential equations texts have a discussion of the slope field, usually called the *direction field*. We prefer the terminology of [BDH] which distinguishes the slope field of a first-order equation from the direction field of an autonomous system. See Section 3.1 for a discussion of vector fields and direction fields.)

When the slope field is not too intricate (that is, when the value of F does not change too fast over small areas of the domain), graphical solutions to the differential equation can be seen in the slope field by squinting one's eyes and letting the line segments run together. The fact that there are infinitely many solution curves visible corresponds to the presence of a constant of integration in the analytic solution to the equation. Note that each line segment of the slope field is actually a tangent line to the graph of one of the solutions.

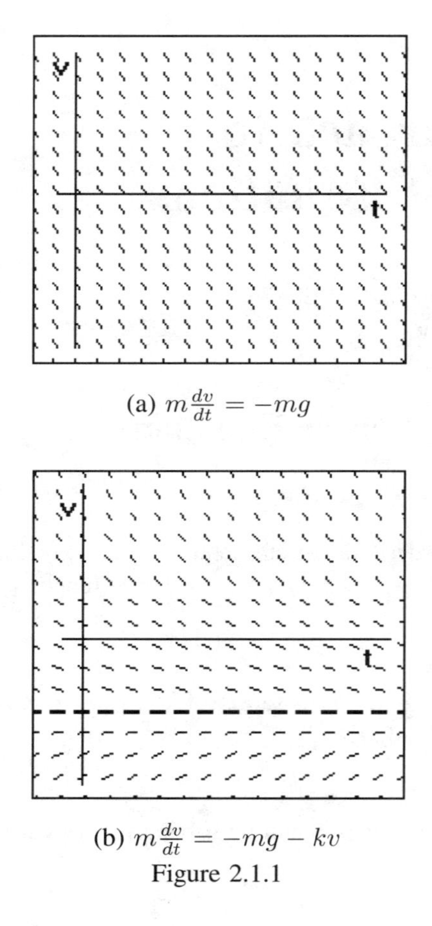

(a) $m\frac{dv}{dt} = -mg$

(b) $m\frac{dv}{dt} = -mg - kv$
Figure 2.1.1

Let us consider two equations for free fall near the earth's surface. The simplest model, often encountered first in a calculus course, is

$$m\frac{dv}{dt} = -mg,$$

where the unknown function is the vertical velocity v, dependent on time t, and the parameters are the mass m and the acceleration g due to gravity. This model ignores air resistance, which is incorporated as a linear term in the next model,

$$m\frac{dv}{dt} = -mg - kv.$$

The parameter $k > 0$ is called the coefficient of drag.

The slope fields for these equations appear in Figure 2.1.1. They represent the function $F(t, v) = -g$ and $F(t, v) = -g - \frac{k}{m}v$, respectively. Notice how they differ, and how easy it is to get a "feel" for the qualitative aspects of a model by looking at its slope field. The model in Figure (a) predicts a steadily decreasing velocity function (increasing speed in the negative direction), while the model in Figure (b) shows a limiting value for the velocity function. This asymptote is defined by the heavier line segments having zero slope, and also represents one of the solutions (the constant solution). This corresponds to the observation that falling objects achieve a terminal velocity.

Some questions to ask of the slope field are:

1. If a ball is thrown upward with initial velocity 100 ft/sec, what will the graph of the velocity function look like? What is the eventual velocity?
2. If a parachutist jumps out of an airplane, what will his velocity function look like?
3. If the above parachutist has been falling at terminal velocity for a few seconds and then opens his parachute, what will his velocity function look like: from the opening of the chute? from the time he left the airplane?
4. How do the values of m and k affect the slope field?

Question 1 may be answered graphically with the two models as follows in Figure 2.1.2.

Thus, the slope field can answer many questions about a differential equation. Producing a slope field is one thing that a computer can do easily, and which is not

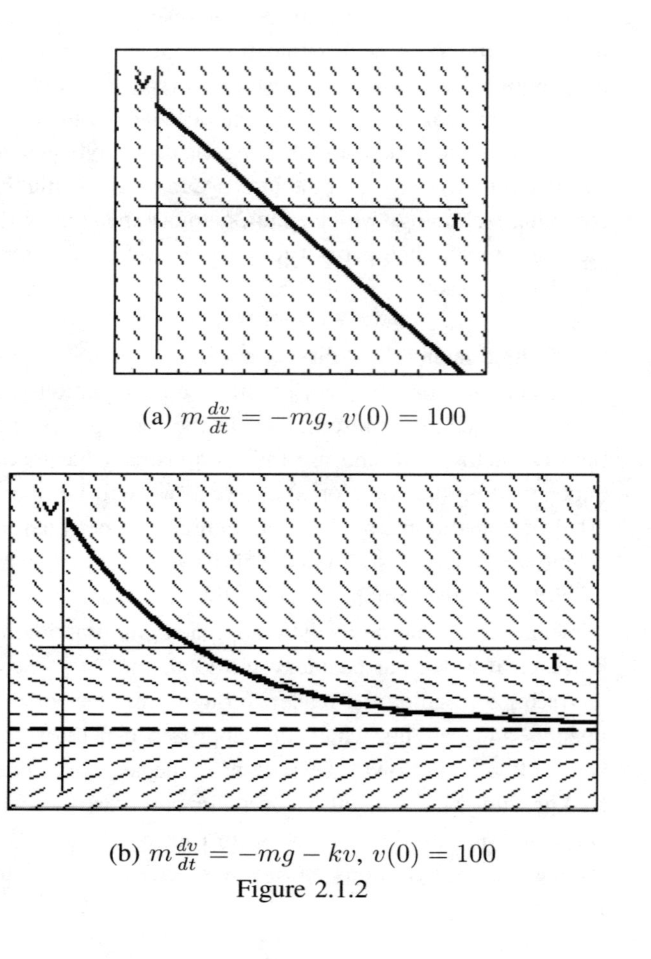

(a) $m\frac{dv}{dt} = -mg,\ v(0) = 100$

(b) $m\frac{dv}{dt} = -mg - kv,\ v(0) = 100$
Figure 2.1.2

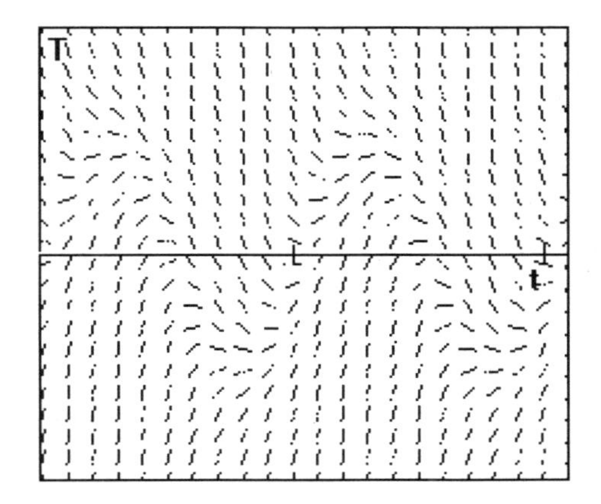

Figure 2.1.3

subject to pesky round-off errors: it is only qualitative information which is wanted. Furthermore, if there is uncertainty about the applicability of the equation to begin with, the slope field can give some preliminary feedback. If the slope field shows the features desired of a model, the more laborious task of finding exact or numerical solutions can be undertaken with confidence.

For another example, consider an object immersed in some medium. Its temperature is governed by Newton's law of cooling: the rate of change of the temperature of the object is proportional to the difference in temperature between the medium and the object. Or, in mathematical language,

$$\frac{dT}{dt} = k(M - T), \quad k > 0,$$

where T is the temperature of the object, t is time, and M is the temperature of the surrounding medium. If M is constant, the equation is the same type as that of free fall: solutions tend to the horizontal asymptote $T = M$. If, however, one postulates a sine wave for M, we obtain a linear nonhomogeneous equation. The slope field has much to tell (see Figure 2.1.3, where $k = 1$ and $M = \sin t$).

The oscillating asymptotic solution is evident, and all of the line segments seem to be directed toward it. The tick marks in the picture are at 2π and 4π; the solution is slightly out of phase with the sine wave of M.

2.2 Autonomous Equations and the Phase Line

An **autonomous** differential equation is one in which the independent variable does not appear. Thus it will have the form $\frac{dy}{dt} = F(y)$. The slope field of such an equation has the property that for a fixed value of y, the slope at points (t, y) is constant. That is, the *isoclines* (curves on

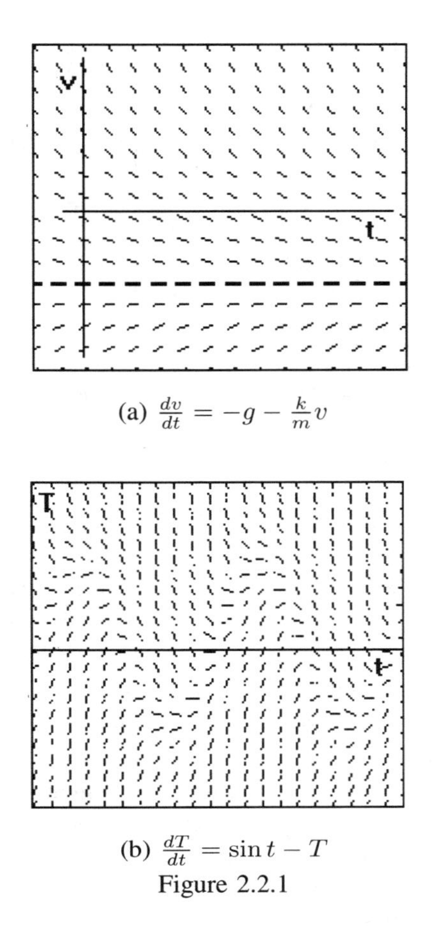

(a) $\frac{dv}{dt} = -g - \frac{k}{m}v$

(b) $\frac{dT}{dt} = \sin t - T$

Figure 2.2.1

which the slope is constant) are horizontal lines. Compare, for example, two slope fields from the previous section given in Figure 2.2.1. The first is autonomous, the second is not.

Autonomous equations model situations in which the rate of change of the system is dependent only upon the current state of the system. Such is the case in the velocity of free fall, heating and cooling in a constant temperature medium, and in many population models. These particular cases have in common that from any given state, the system is "attempting" to achieve some constant state, *equilibrium.* We discuss equilibria further in the next section.

Because the independent variable plays no part in determining the slope of a solution, the slope field can be greatly simplified to show the relevant information and to eliminate the redundancy. We "collapse out" the horizontal axis, and show only the vertical axis. This is called the **phase line** (for first-order equations) or more generally, **phase space** (for nth-order equations). Here is the phase line for free fall next to the slope field:

Algebraically, the phase line is easy to construct: place dots where $\frac{dv}{dt}$ equals zero, and use arrows to indicate the

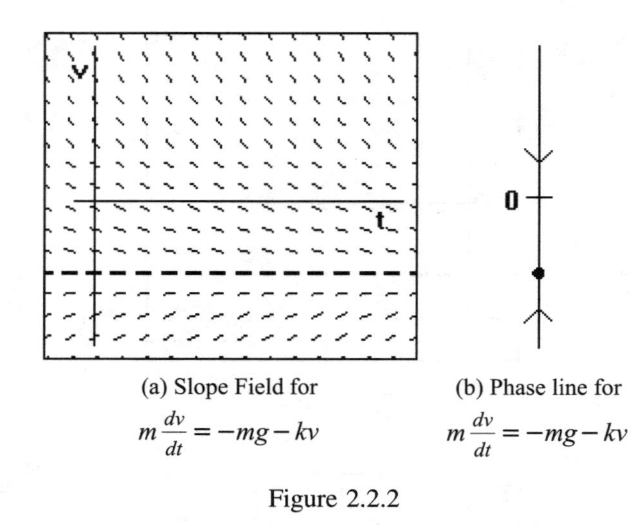

(a) Slope Field for $m\frac{dv}{dt} = -mg - kv$ (b) Phase line for $m\frac{dv}{dt} = -mg - kv$

Figure 2.2.2

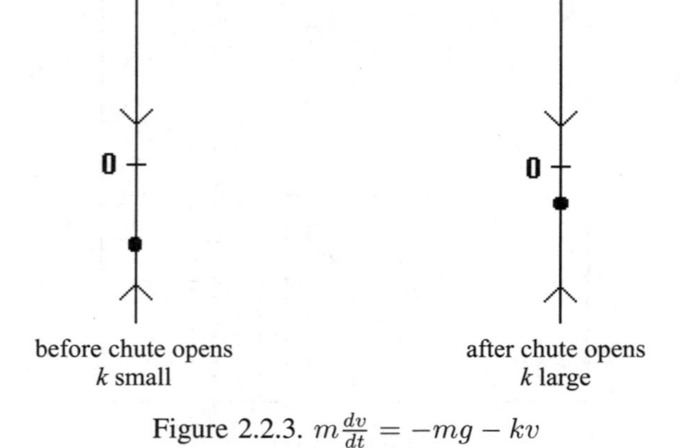

Figure 2.2.3. $m\frac{dv}{dt} = -mg - kv$

sign of $\frac{dv}{dt}$ in between the zeroes. Solutions are obtained from the phase line as follows. If at any time v is equal to the value shown by a dot, subsequent values will not change. This is an equilibrium (constant) solution. If at any time v is within an interval marked with an "up" arrow, subsequent values will increase, either toward an equilibrium if one exists above v, or without bound otherwise. If at any time v is within an interval marked by a "down" arrow, subsequent values will decrease, either toward an equilibrium if one exists below v, or without bound otherwise.

Notice what the phase line for free-fall tells us (Figure 2.2.2b). First, there is one equilibrium solution (terminal velocity). We see that this equilibrium is negative (objects fall downwards!). Second, any velocity larger than equilibrium will "decay" toward equilibrium. Thus, an object travelling upward must come down, and a slowly falling object must speed up. Finally, any velocity smaller than equilibrium will "decay" upward. This means that if an object is travelling downward too fast, it will slow. This is what the parachutist counts on when he opens the chute. Terminal velocity after the chute opens is slower (closer to zero) than before the chute deployed. See Figure 2.2.3.

Learning to draw the phase line can open the door to important theoretical questions:

1. Why is there only one direction for an arrow to point in between the equilibria?
2. Does this mean that solutions to autonomous first-order equations cannot oscillate?
3. Can a solution increasing toward an equilibrium ever reach the equilibrium?
4. Does a solution increasing without bound do so for all time? (The alternative would be for a solution to have a vertical asymptote.)

For another example, consider the familiar population model

$$\frac{dP}{dt} = aP(M-P) - h$$

known as logistic growth with harvesting. The variables are P for population and t for time. As for the parameters, M is the theoretical equilibrium population, a is the growth constant, and h is the harvesting rate. The equation is autonomous and nonlinear. (It can be solved by separation of variables and partial fractions, if you must know.) Let us look at the geometry of the phase line.

If $h = 0$, the equilibria are at $P = 0$ and $P = M$. For small h, the equilibria are found by solving the quadratic equation $aP(M-P) - h = 0$, whose solutions are

$$P = \frac{M}{2} \pm \frac{1}{\sqrt{a}}\sqrt{\frac{aM^2}{4} - h}.$$

And for $h > \frac{aM^2}{4}$, there are no equilibria. It is easily seen that between the equilibria $\frac{dP}{dt} > 0$ and $\frac{dP}{dt}$ is negative elsewhere. Several representative phase lines are shown in Figure 2.2.4.

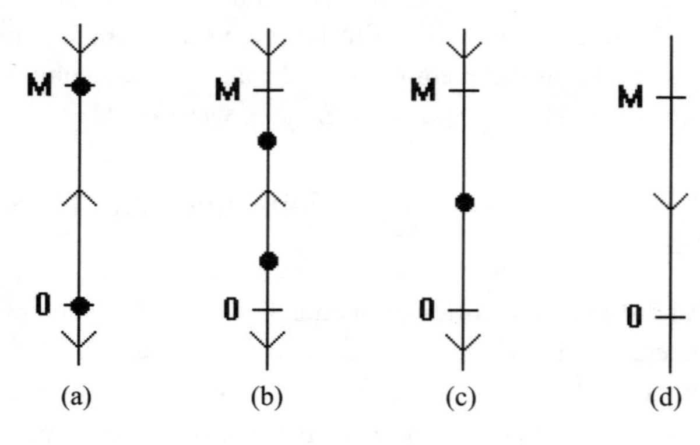

Figure 2.2.4. $\frac{dP}{dt} = aP(M-P) - h$

The first (a) shows no harvesting. We see the expected equilibria of $P = 0$ and $P = M$. Note that for any population less than M, the population grows toward M, and for any population over M, the population decreases toward M. It would seem that this model represents a population having an ideal value supported by its environment.

As harvesting increases (b), the phase line clearly shows the effect on the population. For small h, there is still an ideal equilibrium, but it is at a lower level than before harvesting. Further, there is now another positive equilibrium below which this population should not go. If it does, it must decrease eventually to zero. For large h (d), there is no equilibrium; any initial population is doomed to die out. The value $h = \frac{aM^2}{4}$ (c) seems to be a critical value. There is more on the changing nature of the phase line under the effects of parameters in the next section.

2.3 Fixed Points, Stability, and Bifurcation

The logistic population model with harvesting,

$$\frac{dP}{dt} = aP(M - P) - h,$$

discussed in the last section, is an excellent introduction to the ideas of this section: *fixed points* for a differential equation, *stability* of a fixed point, and *bifurcation* values of *parameters*. Here are the definitions:

A **fixed point** for a differential equation is a constant solution, also called an equilibrium. Thus, a fixed point is a value of the dependent variable which makes the slope zero for all values of the independent variable. Fixed points are therefore most frequently discussed in the context of autonomous equations.

A fixed point p is called **stable** if any solution with value close enough to p has subsequent values approaching p. Some books call a stable point a *sink* or an *attractor*.

A **parameter** is any quantity in a differential equation which is constant. In the logistic equation above, the parameters are a, M, and h, whereas the variables are P and t. In the contrary nature of mathematical nomenclature, as soon as one labels a constant a "parameter," one is usually preparing to change its value.

Thus, a **bifurcation** occurs at a parameter value c if the qualitative behavior of the differential equation changes as the parameter increases through c.

Exercise 1: For which values of h (in terms of a and M) does the logistic equation given above have fixed points? Which fixed points are stable? Find the bifurcation value of h (in terms of a and M) and explain what kind of qualitative change occurs in the equation at this value. (If you turned back to Figure 2.2.4 in the previous section to answer these questions, you get an "A.")

As an application, suppose there is a resort area which stocks its lake with fish. We measure the fish in tons and the time in months. Through the work of biologists, we ascertain that the logistic equation applies to the fish population living and reproducing in the lake: $\frac{dP}{dt} = 0.004P(1000 - P)$. We see theoretically that if we allow fishing of h tons per month, the equation becomes

$$\frac{dP}{dt} = 0.004P(1000 - P) - h,$$

and if $h < \frac{(.004)(1000)^2}{4} = 1000$ tons/month (the bifurcation value), we have a sustainable population of more than 500 tons of fish in the lake. This is a result of the existence of a stable fixed point in the differential equation. In this case, we need not go to the expense of stocking the lake any further.

If the biologists also tell us that some years we can expect a sudden fish kill of about 200 tons due to overgrowth of algae, we must allow for a margin of safety. We choose a value of h for which the stable fixed point is at least 200 units above the unstable fixed point. This is left as Exercise 2. Figure 2.3.1 gives the full story.

There is another graphical device which can help us to picture the phase line of an autonomous equation. That is to graph the function defining the equation. For example, for the logistic equation above, one would graph $F(P) = 0.004P(1000 - P) - h$ (Figure 2.3.2). This shows where the fixed points are (zeroes of the equation) and which are stable. Consider the case when h is zero. Remember that $F(P)$ represents the derivative of the solution at the value P.

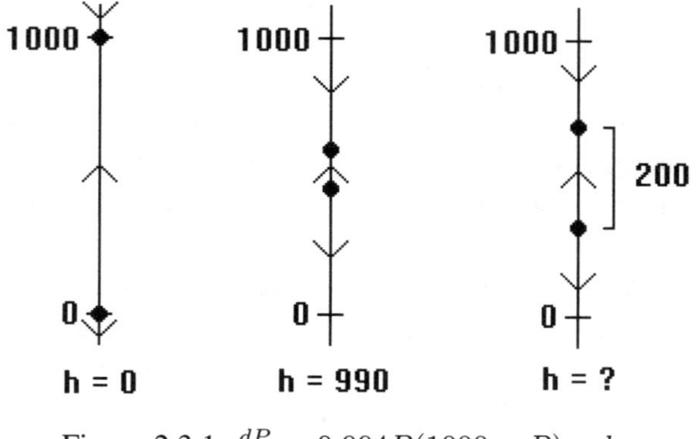

Figure 2.3.1. $\frac{dP}{dt} = 0.004P(1000 - P) - h$

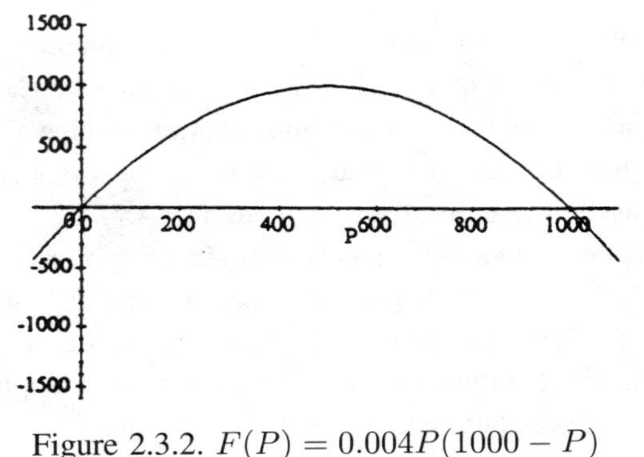

Figure 2.3.2. $F(P) = 0.004P(1000 - P)$

The stable fixed point $P_0 = 1000$ has the following properties. First $F(1000) = 0$. Second, for initial P value below 1000, solutions increase (that is, $F(P) > 0$). And third, for initial P values above 1000, solutions decrease ($F(P) < 0$). Thus the function F must be decreasing through $P = 1000$, so $F^{'}(1000) < 0$. The mathematical conditions $F(P_0) = 0$ and $F^{'}(P_0) < 0$ signify a stable fixed point at $P = P_0$. (A stable fixed point may satisfy $F^{'}(P_0) = 0$, but this condition alone is inconclusive.)

Graphing F for several values of h also shows us where and why a bifurcation occurs:

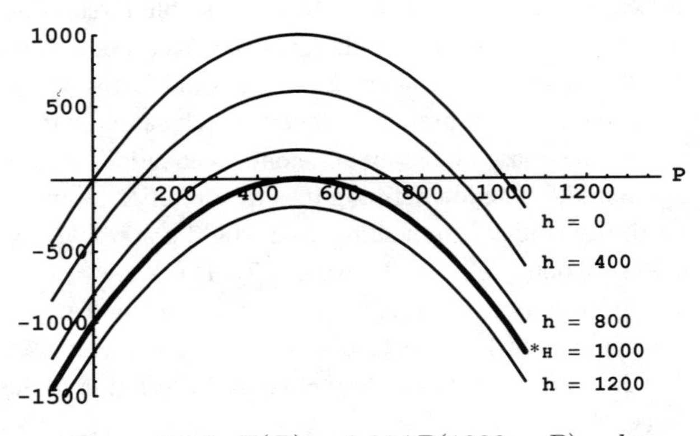

Figure 2.3.3. $F(P) = 0.004P(1000 - P) - h$

There are different kinds of bifurcations. The bifurcation at $h = \frac{aM^2}{4}$ in the logistic equation with harvesting (above) is one in which two fixed points, one stable, the other not, coalesce and disappear. The situation is shown graphically in the *bifurcation diagram*, a graph of P vs. h. The idea is to line up all the phase lines for the differential equation side by side, placed according to their corresponding values of h. See Figure 2.3.4. The heavy curve shows the fixed points. As h increases through $\frac{aM^2}{4}$, two fixed points become one, and then disappear.

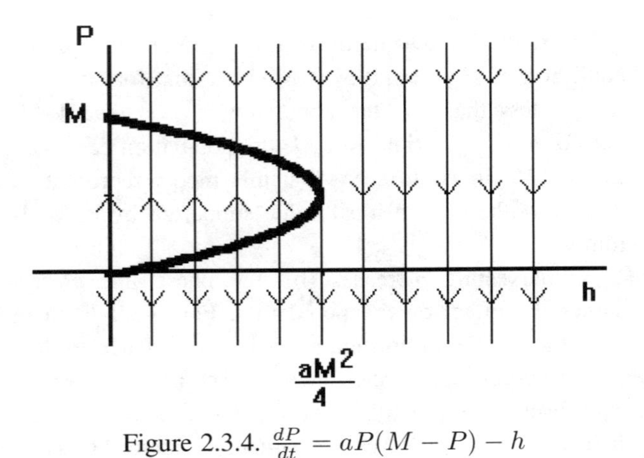

Figure 2.3.4. $\frac{dP}{dt} = aP(M - P) - h$

Exercise 3: Draw the bifurcation diagram for the differential equation

$$\frac{dy}{dt} = y^3 + cy.$$

This bifurcation is called a *pitchfork* bifurcation.

Answers to Exercises

1. There are fixed points for $0 \leq h \leq \frac{aM^2}{4}$. If $h < \frac{aM^2}{4}$, the larger one is an attractor and the smaller is called a *repeller*. The bifurcation value is $\frac{aM^2}{4}$. Below this value, the differential equation has two fixed points. At $h = \frac{aM^2}{4}$, there is only one fixed point, unstable. Above $\frac{aM^2}{4}$, there are no fixed points. In short, two fixed points become one, then disappear.
2. The value $h = 960$ gives fixed points 400 and 600, 200 units apart.
3. The bifurcation value is at $c = 0$. When $c < 0$, there are three fixed points, $-\sqrt{-c}$, 0, and $\sqrt{-c}$. The value 0 is an attractor and the other two fixed points are repellers. At $c = 0$, the three fixed points coalesce into one unstable fixed point, a repeller: see Figures 2.3.5 and 2.3.6.

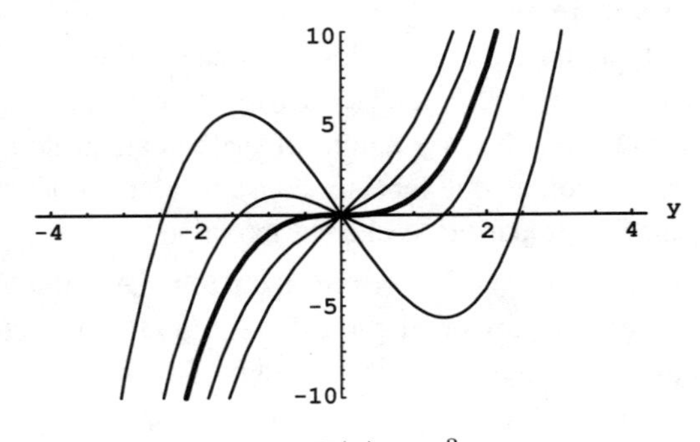

Figure 2.3.5. $F(y) = y^3 + cy$

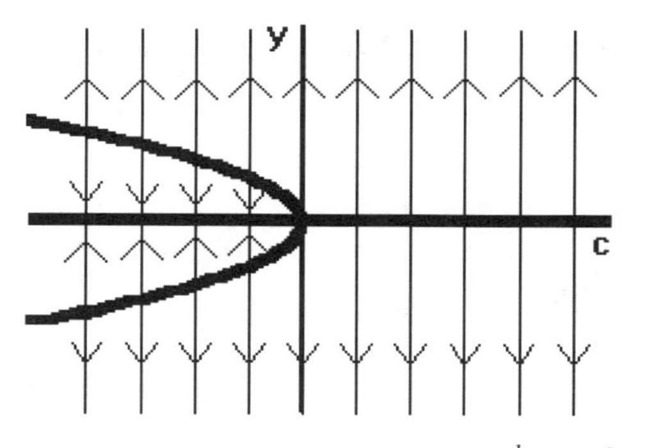

Figure 2.3.6. Bifurcation Diagram for $\frac{dy}{dt} = y^3 + cy$

2.4 The Potential Function

Another way to determine the qualitative behavior of an autonomous differential equation $\frac{dy}{dt} = F(y)$ is to use the **potential function** for F.

Definition: P is a **potential function** for F iff $P'(y) = -F(y)$.

Consider the example

$$\frac{dy}{dt} = y^2 - 4$$

A potential function is $P(y) = 4y - \frac{y^3}{3}$.

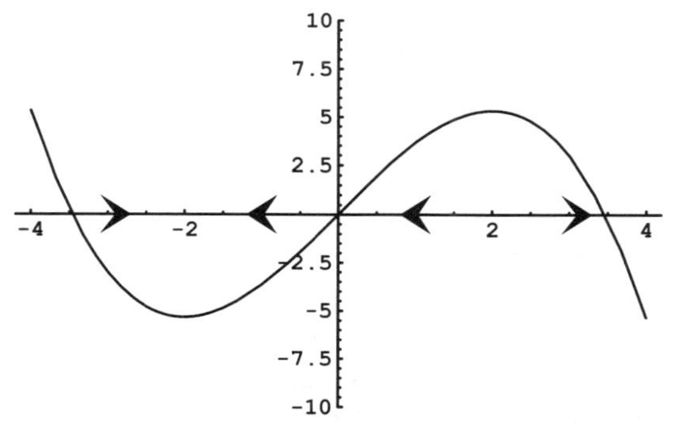

Figure 2.4.1

One can see that stable fixed points correspond to **relative minima** of the potential function, and unstable fixed points correspond to **relative maxima** of the potential function.

Now consider

$$\frac{dy}{dt} = y^4 - 5y^3 + 8y^2 - 4y$$

One can factor this to see that

$$f(y) = y(y-1)(y-2)^2$$

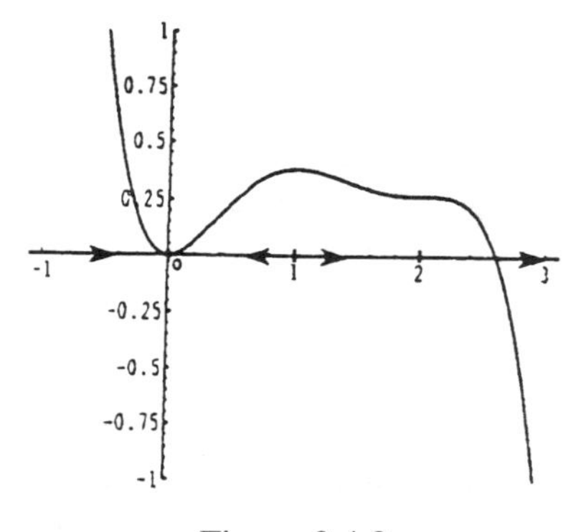

Figure 2.4.2

The equilibria are then 0, 1 and 2. A potential function is

$$P(y) = \frac{-y^5}{5} + \frac{5y^4}{4} - \frac{8y^3}{3} + \frac{4y^2}{2}$$

One can see that 0 is an attractor, 1 is a repeller, and 2 is neither.

We can summarize the use of the potential function as follows:

Theorem *Let* $\frac{dy}{dt} = F(y)$ *be a first order, autonomous ordinary differential equation, and* F *have* **isolated** *zeros. Let* P *be a potential function for* F *(i.e.,* $P' = -F$*.)* P *has a critical point at* y_0 *iff* y_0 *is a fixed point,* y_0 *is a relative max. of* P *iff* y_0 *is a repeller, and* y_0 *is a relative min. of* P *iff* y_0 *is an attractor.*

3. Systems of Equations

3.1 Vector Fields, Direction Fields, and Solution Curves

One geometric way to look at a system of differential equations is as a **vector field**; that is, a function whose domain is an open set in Euclidean space and whose range is a set of vectors anchored at those points. Consider the familiar system which represents the undamped motion of a mass attached to a linear spring:

$$\begin{cases} \frac{dx}{dt} = y \\ \frac{dy}{dt} = -x \end{cases}$$

where x is the vertical displacement of the mass from equilibrium and y is the vertical velocity. The vector field is the function $F : \mathbb{R}^3 \longrightarrow \mathbb{R}^3$ given by

$$F(t, x, y) = [1, y, -x] \ ,$$

where at time t, position x, and velocity y, those values are changing at rates 1, y, and $-x$, respectively. We may

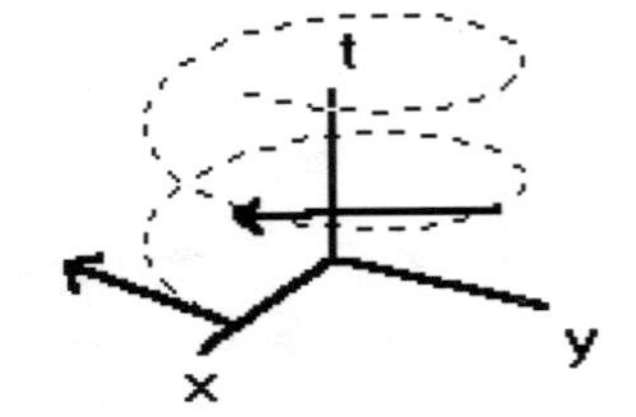

Figure 3.1.1. $F(t, x, y) = [1, y, -x]$

graph this function, similarly to the way we graph the slope field for a single equation. The graph exists in 3-space, and is a collection of vectors. The vector $F(t, x, y)$ is placed with its tail at the point (t, x, y). See Figure 3.1.1.

The two vectors shown are $[1, 0, -1] = F(0, 1, 0)$ originating at the point $(0, 1, 0)$, and $[1, 1, 0] = F(\frac{3\pi}{2}, 0, 1)$ originating at the point $(\frac{3\pi}{2}, 0, 1)$. The dotted line is the graph of the parametric solution curve $(t, x(t), y(t))$ which corresponds to the initial condition $x(0) = 1$, $y(0) = 0$. The vectors are tangent to this solution curve.

As with the slope field, considering the graph of the vector field can answer qualitative questions about a system of differential equations. For example, it appears that the x- and y-coordinates of the above vector field are bounded, and that their values repeat over time. Perhaps the solutions are periodic.

Collapsing out each of the three axes in the above graph gives other familiar graphs. For example, if just the t and x coordinates are graphed, one obtains the position vs. time graph, in this case, the particular solution $x = \cos t$. See Figure 3.1.2, and compare this graph to the 3-dimensional version above.

Collapsing out the t -axis is particularly revealing in the case where the system is autonomous. This is an n-dimensional analogue of the phase line, and is the topic of the next section.

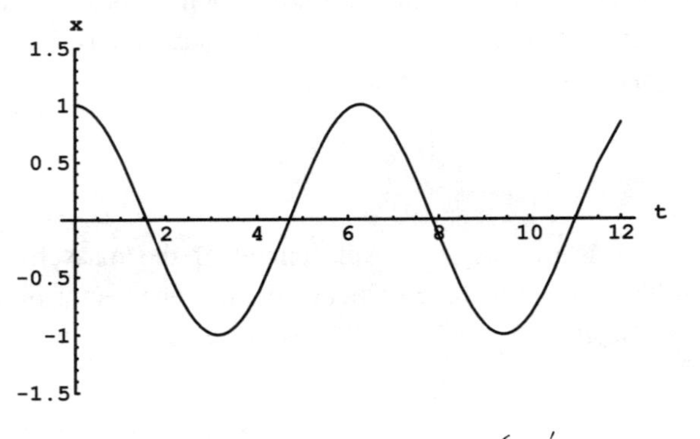

Figure 3.1.2. x vs. t from the system $\begin{cases} x' &= y \\ y' &= -x \end{cases}$

3.2 Autonomous Equations and the Phase Space

An autonomous system of differential equations is a system whose equations do not contain the independent variable. The spring-mass system of the previous section is such a system:

$$\begin{cases} \frac{dx}{dt} &= y \\ \frac{dy}{dt} &= -x \end{cases}$$

The vector field of an autonomous system has vectors which are independent of the value of t, and thus the portion of the vector field containing just the dependent variable information is very useful. Look at Figure 3.1.1 in the previous section, and collapse out the t-axis. What do you see? The solution curve should appear now as a circle in the xy-plane, and the vectors are tangent to it.

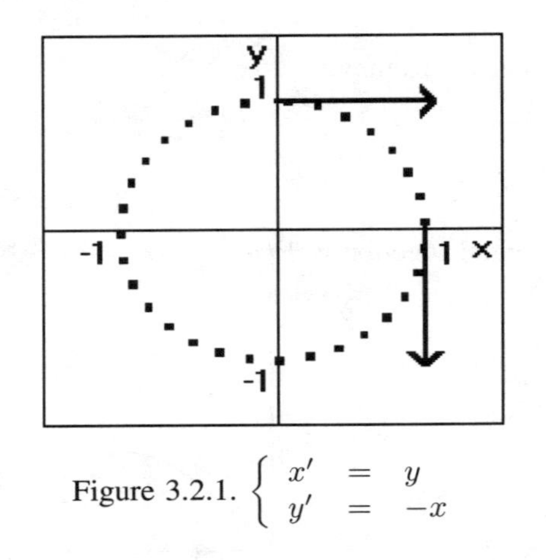

Figure 3.2.1. $\begin{cases} x' &= y \\ y' &= -x \end{cases}$

Given an autonomous system of n differential equations in n unknown functions, the n-dimensional Euclidean space containing axes representing the dependent variables is called the *phase space* of the system. As with the phase line, the vector field represented in phase space provides useful qualitative information about solutions to the system. Further, it is easily generated by a computer, without making use of the actual solutions. Figure 3.2.2 is a more complete vector field for the spring-mass system, suggesting solutions for more initial conditions.

What can we learn about the system of equations from this picture? The directions of the vectors seem to indicate circular solutions. (Closed curve solutions correspond to periodicity.) Some of the circles have small radii, indicating solutions of small amplitude. Other solutions have larger amplitude, and these also have longer vectors attached to them. This means that larger oscillations of the mass occur at higher speeds. One of the vectors, at the

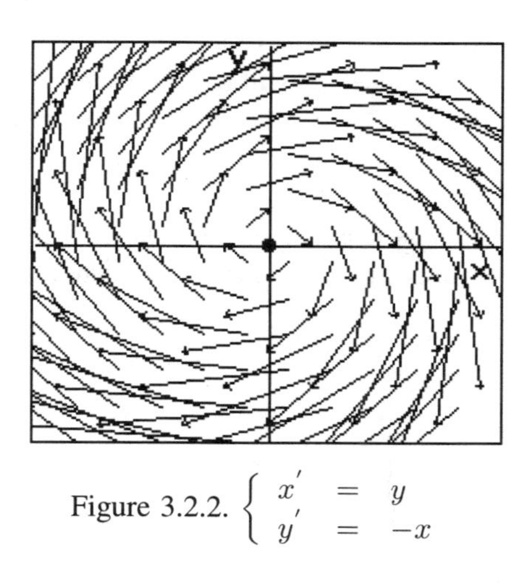

Figure 3.2.2. $\begin{cases} x' &= y \\ y' &= -x \end{cases}$

origin, has zero length. This must be a solution in which both x and y are zero and never change, i.e., an equilibrium solution, or fixed point. There is more on fixed points in the following section.

Notice that the vector field can get pretty messy for at least two reasons. The first is that the vectors can have large magnitude. The second is that their directions may change rapidly as x and y change, and so one is tempted to draw many vectors in a small area. The problem of long vectors can be solved by drawing the **direction field**: the vector field with vectors normalized to unit length. This still provides much qualitative information about direction, periodicity, fixed points, and stability, leaving out only the speed. See Figure 3.2.3.

Here are two more examples of direction fields. They represent different models of two populations interacting in a predator-prey relationship. The first is the Lotka-Volterra system. The second is similar, with a crowding term added for both species.

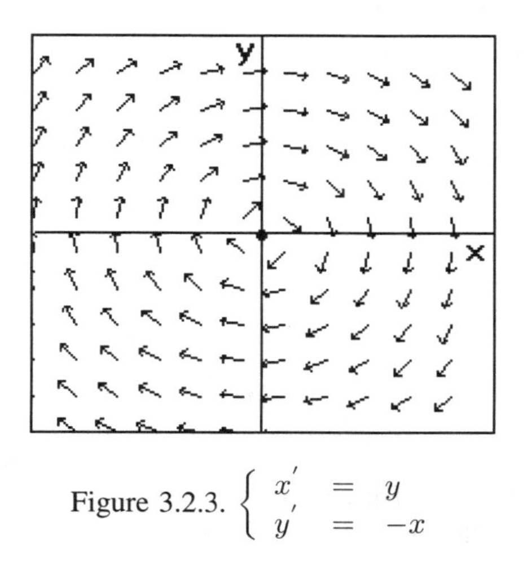

Figure 3.2.3. $\begin{cases} x' &= y \\ y' &= -x \end{cases}$

$$\begin{cases} x' &= x - xy \\ y' &= -y + 2xy \end{cases}$$

Figure 3.2.4 (a)

Figure 3.2.4a is a direction field strongly suggestive of periodic solutions. Both prey and predator experience cyclic growth and decline. Can you see the equilibrium solution at the center of the cycles? You can find both equilibria (only one nontrivial one) by setting both equations equal to zero and solving for x and y.

There is something a little different in Figure 3.2.4b. Notice the directions of vectors near the x-axis. One wants to focus on the area near the lower left corner of the first quadrant. An enlargement (Figure 3.2.5a) suggests that solutions spiral into the equilibrium, rather than cycling periodically. Some numerically generated solutions are shown on top of the original direction field. See Figure 3.2.5b. This system has four equilibria, one nontrivial. They can be easily found algebraically.

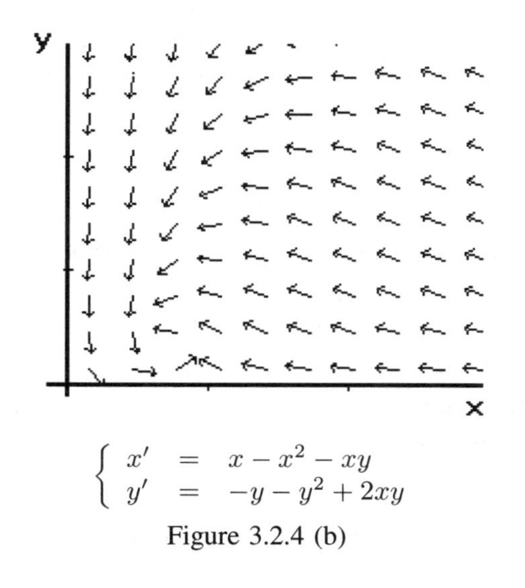

$$\begin{cases} x' &= x - x^2 - xy \\ y' &= -y - y^2 + 2xy \end{cases}$$

Figure 3.2.4 (b)

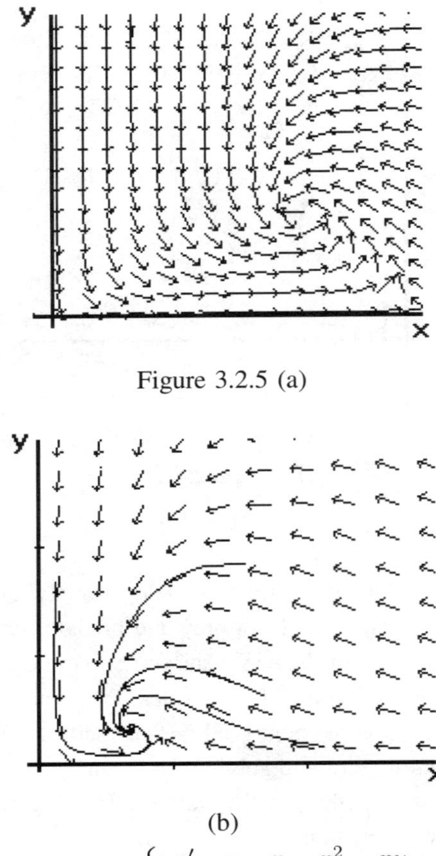

Figure 3.2.5 (a)

(b)

Figure 3.2.5. $\begin{cases} x' &= x - x^2 - xy \\ y' &= -y - y^2 + 2xy \end{cases}$

3.3 Fixed Points, Stability, and Bifurcation

There is much to say about stability in systems of differential equations. The linear case has been the most extensively studied. Necessarily, this article will only touch on some fundamental concepts.

The definitions of **fixed point** and **stability** are the same as for single equations (see Section 2.3 above), if one understands by a *solution* a vector solution

$$X = \begin{bmatrix} x_1 \\ x_2 \\ \vdots \end{bmatrix}.$$

We will make increasing use of the vector notation

$$\frac{dX}{dt} = F(t, X),$$

where $X : \mathbb{R} \longrightarrow \mathbb{R}^n$ is the unknown vector of solution functions and $F : \mathbb{R}^{n+1} \longrightarrow \mathbb{R}^n$. In the case that the system is autonomous, $F : \mathbb{R}^n \longrightarrow \mathbb{R}^n$. If the system is linear, then we will write

$$\frac{dX}{dt} = F(t, X) = A(t)X + E(t),$$

where A is an $n \times n$ matrix and E is an $n \times 1$ matrix. In all of our examples, A will be a constant 2×2 matrix and $E \equiv 0$. Even this restricted case holds useful information for more general systems. There is more on this connection in Section 3.4 below.

As an example, the undamped linear spring-mass system of the previous section may be written

$$\frac{dX}{dt} = \begin{bmatrix} 0 & 1 \\ -1 & 0 \end{bmatrix} X,$$

where $X = \begin{bmatrix} x \\ y \end{bmatrix}$. The unique fixed point is $X = \begin{bmatrix} 0 \\ 0 \end{bmatrix}$. This fixed point is in the gray area between stable and unstable, as nearby solutions neither approach it asymptotically nor tend away from it. It is called a **center**. See the direction field of Figure 3.2.3 in the previous section. The exact solution to this system is

$$X = \begin{bmatrix} C_1 \sin t + C_2 \cos t \\ C_1 \cos t - C_2 \sin t \end{bmatrix},$$

and the periodic nature of the motion is evident.

What might a bifurcation in a 2-dimensional system look like? Let us add linear damping to the spring-mass equation above. This will add a term to the acceleration, $y' = -x - ky$ ($k > 0$), and a parameter to the system:

$$\frac{dX}{dt} = \begin{bmatrix} 0 & 1 \\ -1 & -k \end{bmatrix} X.$$

When $k > 0$, we expect the oscillations to decrease in both amplitude and speed. Thus, we should have a bifurcation at $k = 0$: the fixed point should become stable. For $k < 0$, the solutions tend away from the fixed point. You may want to invent a physical interpretation for $k < 0$; the authors will not venture one. The direction fields of these three cases are shown in Figure 3.3.1.

Sample solutions to the above equations are

$$X = e^{\frac{1}{2}t} \begin{bmatrix} C_1 \sin \frac{\sqrt{3}}{2}t + C_2 \cos \frac{\sqrt{3}}{2}t \\ (-\frac{\sqrt{3}}{2}C_2 + \frac{1}{2}C_1) \sin \frac{\sqrt{3}}{2}t \\ +(\frac{\sqrt{3}}{2}C_1 + \frac{1}{2}C_2) \cos \frac{\sqrt{3}}{2}t \end{bmatrix}, \quad k = -1$$

and

$$X = e^{-\frac{1}{2}t} \begin{bmatrix} C_1 \sin \frac{\sqrt{3}}{2}t + C_2 \cos \frac{\sqrt{3}}{2}t \\ (-\frac{\sqrt{3}}{2}C_2 - \frac{1}{2}C_1) \sin \frac{\sqrt{3}}{2}t \\ +(\frac{\sqrt{3}}{2}C_1 - \frac{1}{2}C_2) \cos \frac{\sqrt{3}}{2}t \end{bmatrix}, \quad k = 1$$

You did ask? The analytic solutions verify the spiraling behavior: the exponential terms either blow up ($k = -1$)

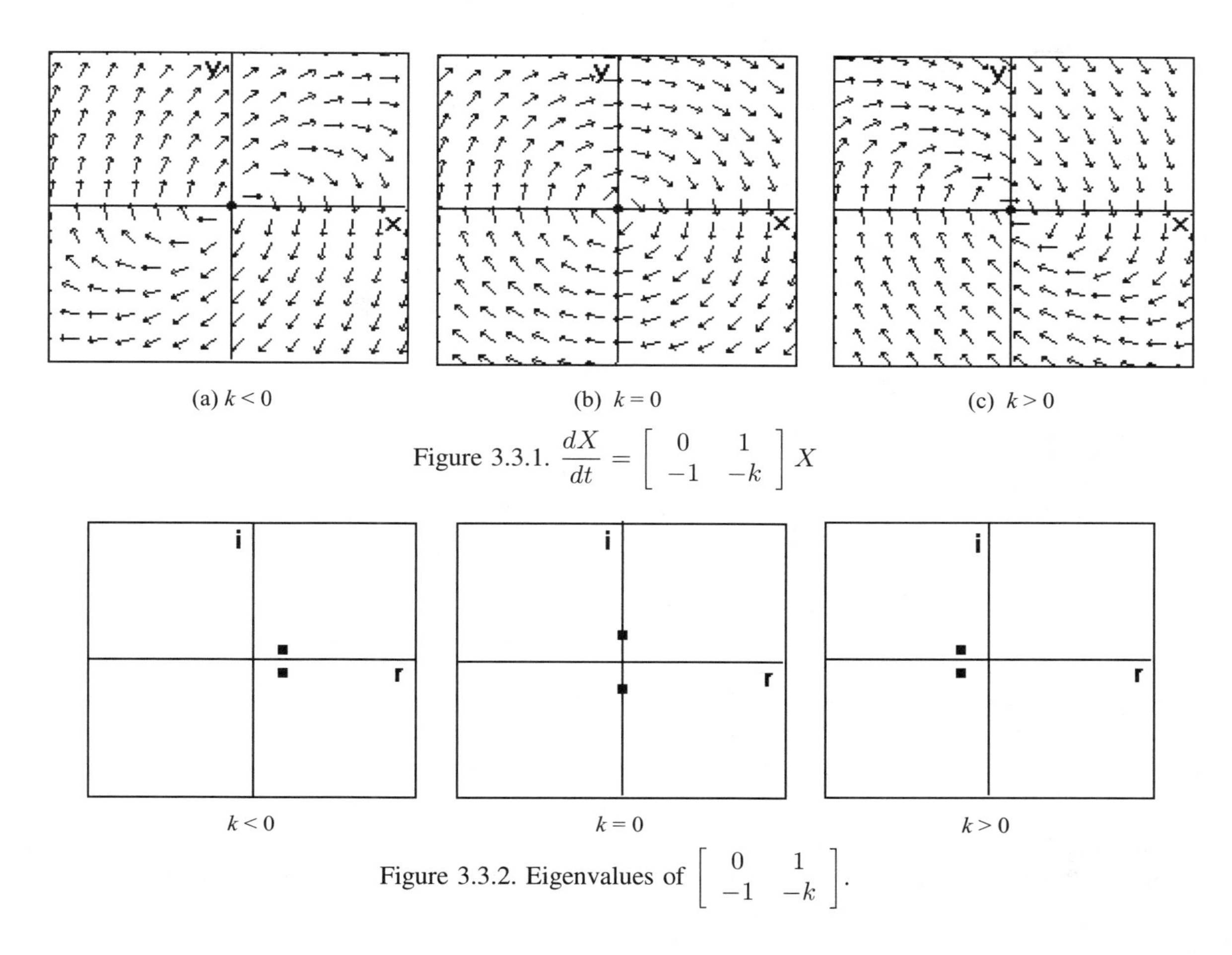

(a) $k<0$ (b) $k=0$ (c) $k>0$

Figure 3.3.1. $\dfrac{dX}{dt} = \begin{bmatrix} 0 & 1 \\ -1 & -k \end{bmatrix} X$

$k<0$ $k=0$ $k>0$

Figure 3.3.2. Eigenvalues of $\begin{bmatrix} 0 & 1 \\ -1 & -k \end{bmatrix}$.

or decay ($k = 1$). This bifurcation is one in which the nature of the single fixed point changes, but the number and value of fixed points does not.

The theory of linear systems of differential equations tells us that the nature of the solutions (and of the fixed point) depends on the eigenvalues of the matrix A. (See, for example, the discussion in Chapter 7 of [BD].) Considering the matrix for the above system,

$$A = \begin{bmatrix} 0 & 1 \\ -1 & -k \end{bmatrix},$$

we compute the eigenvalues to be

$$\lambda = \frac{-k \pm \sqrt{(k^2-4)}}{2} = -\frac{k}{2} \pm \sqrt{\left(\frac{k}{2}\right)^2 - 1}.$$

Now we can see a somewhat more interesting story. The above remarks about the solutions apply only to oscillating solutions, resulting from k-values between -2 and 2. For these values, λ is complex; the real part of λ determines the stability of the solution. When $0 < k < 2$, the real part of λ is negative, making [0,0] an attractor. (See the above solution for $k = 1$. For this parameter value, $\lambda = -\frac{1}{2} \pm \frac{\sqrt{3}}{2} i$. The real part, $-\frac{1}{2}$, creates the exponential term $e^{-\frac{1}{2}t}$, indicating decay of the motion, or stability.) Similarly, when $-2 < k < 0$, the real part of λ is positive, indicating a repeller. And when $k = 0$ the real part of λ is zero, indicating a center. Figure 3.3.2 shows the eigenvalues as k increases through 0.

The eigenvalues remain complex conjugates, changing their real parts from positive to negative.

The algebra suggests a different kind of bifurcation at $k = \pm 2$. At $k = 2$ (large damping), we expect the oscillation of the physical system to give way to a monotone motion toward equilibrium. This would be the case if we transferred the spring and mass from the air into a vat of honey. Figure 3.3.3 shows the eigenvalues in the complex plane as k increases through 2.

The eigenvalues change from conjugate pairs, to a repeated real value, to distinct real values. The reader is invited to look at the bifurcation of eigenvalues as k increases through -2.

Here is one more quick example to show the birth of fixed points. The system is nonlinear:

$$\frac{dX}{dt} = \begin{bmatrix} x_2 \\ k - x_1^2 \end{bmatrix}.$$

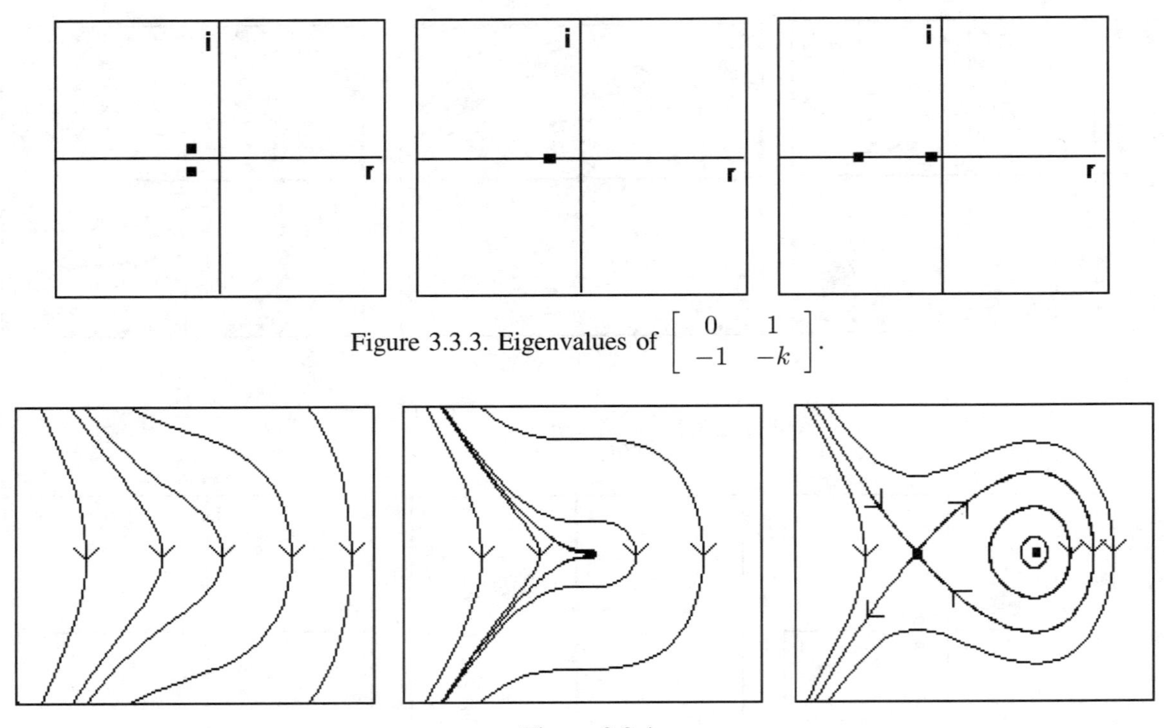

Figure 3.3.3. Eigenvalues of $\begin{bmatrix} 0 & 1 \\ -1 & -k \end{bmatrix}$.

Figure 3.3.4

The sign of k determines whether there are fixed points at all. If $k < 0$ there are none; if $k = 0$ the only fixed point is $X = [0, 0]$; and if $k > 0$ there are two, $X = [\pm\sqrt{k}, 0]$. None are stable, but in the latter case, there is a center. The phase portrait solutions are shown in Figure 3.3.4.

It appears that the solution curves are pinched together to form first one, then two fixed points.

3.4 Linearization

Since it's easy to understand the geometry of a linear system, one would hope that it could also be used to understand that of a non-linear system. For example, consider the second-order ODE

$$y'' - (y')^2 + y^2 - y = 0$$

Let $x_1 = y$ and $x_2 = y'$; then

$$\begin{aligned} x_1' &= x_2 \\ x_2' &= x_2^2 - x_1^2 + x_1 \end{aligned}$$

Notice that these are autonomous equations, one of which is *non-linear*. The equilibria occur when $x_1 = 0, 1$ and $x_2 = 0$. We would like to be able to at least describe the behavior of solutions near these equilibria.

Recall the following theorem from multivariable calculus:

Theorem. *Let $F : R^n \to R$ be a function which is differentiable at X_0. The best affine approximation for F near X_0 is given by*

$$F(X_0) + F'(X_0)(X - X_0).$$

Now in our example we have

$$x_2' = F(x_1, x_2)$$

where F is non-linear. We shall replace F by its best affine approximation. At $x_0 = (0, 0)$ we get

$$\begin{aligned} F(x_1, x_2) &\approx F(0,0) + F'(0,0)((x_1, x_2) - (0,0)) \\ &= 0 + (1,0)(x_1, x_2) \\ &= x_1 \end{aligned}$$

So, near the fixed point $(0, 0)$, we can approximate our non-linear system by the linear system

$$\begin{aligned} x_1' &= x_2 \\ x_2' &= x_1 \end{aligned}$$

The matrix for this linear system is

$$A = \begin{bmatrix} 0 & 1 \\ 1 & 0 \end{bmatrix}$$

This has eigenvalues $\lambda = \pm 1$, so (0,0) is a saddle.

Near the fixed point (1,0), we get the approximation

$$F(x_1, x_2) \approx -x_1 + 1$$

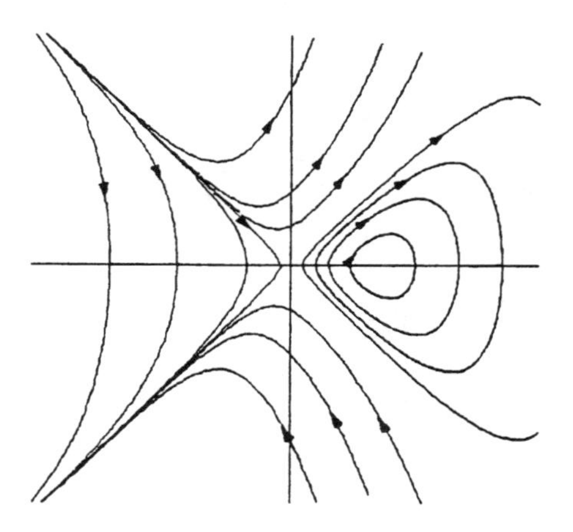

Figure 3.4.1

So, near the fixed point (1,0), we can approximate our non-linear system by the linear system

$$\begin{aligned} x_1' &= x_2 \\ x_2' &= -x_1 + 1 \end{aligned}$$

The matrix for it is

$$A = \begin{bmatrix} 0 & 1 \\ -1 & 0 \end{bmatrix}$$

This has eigenvalues $\lambda = \pm i$, so $(1, 0)$ is a center. Figure 3.4.1 shows the actual phase portrait for the system.

How good is our approximation? In general, we can say the following.

The Hartman-Grobman Theorem. *Suppose the system $\frac{dX}{dt} = F(X)$ is approximated near an equilibrium point, x_0, by the linear system $\frac{dX}{dt} = AX + E$. If A has no eigenvalues which are zero or purely imaginary, then the phase portrait of the original system near x_0 can be obtained from the phase portrait of the linearized system by a continuous transformation.*

As our example shows, the theorem does not preclude an accurate approximation if the hypothesis on the eigenvalues doesn't hold. One can also get a good feel from the non-linear phase portrait above for how the linearized system's phase portrait has been transformed, including its eigenlines.

3.5 Limit Cycles

Suppose we look at an electrical circuit containing a resistor, R, an inductor, L, and a capacitor, C, with current flowing as indicated by the arrows:

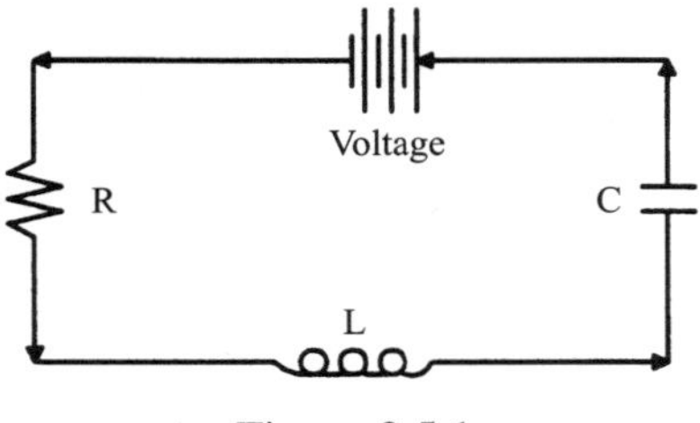

Figure 3.5.1

Using Kirchhoff's Laws, and Faraday's Law we can derive equations for the circuit

$$\begin{aligned} \frac{dx}{dt} &= y - x^3 + \mu x \\ \frac{dy}{dt} &= -x \end{aligned}$$

where x is the current through the inductor, y is the voltage drop across the inductor, and we have assumed a generalized Ohm's Law: voltage drop across the resistor $= x^3 - \mu x$. μ represents a parameter related to the resistor, for example, its temperature (see [HS].) If we look at the phase portraits and the corresponding eigenvalues of this system for three different values of μ we see Figure 3.5.2.

As μ goes from -1 to 1, the eigenvalues go from complex with negative real part to complex with positive real part, but at $\mu = 0$ they are pure imaginary. Thus we see a bifurcation. The phase portrait shows a spiral sink at $(0, 0)$, then a center, then a spiral source. Further, the solution curves near the source are spiraling out to a closed curve that is itself a solution. In fact, one can show that solution curves outside the periodic solution spiral into it as well. Our electrical circuit has gone from one which dies out, to one which achieves a continuing oscillation described by a periodic equilibrium.

Definition: A **limit cycle** is a closed solution curve for which there exist solutions that approach it asymptotically in either forward or backward time.

This sort of bifurcation wherein a limit cycle is formed around an equilibrium point is called a **Hopf Bifurcation.** One of the tools for detecting limit cycles in the plane (there is not an analog in higher dimensions) is the **Poincare-Bendixson Theorem**.

Theorem. *If $\frac{dX}{dt} = F(X)$ where $F : R^2 \to R^2$ is C^1, then any non-empty compact limit set containing no equilibrium point is a limit cycle.*

So if one can show the existence of such a limit set, one knows there is a limit cycle to be had. Consider the

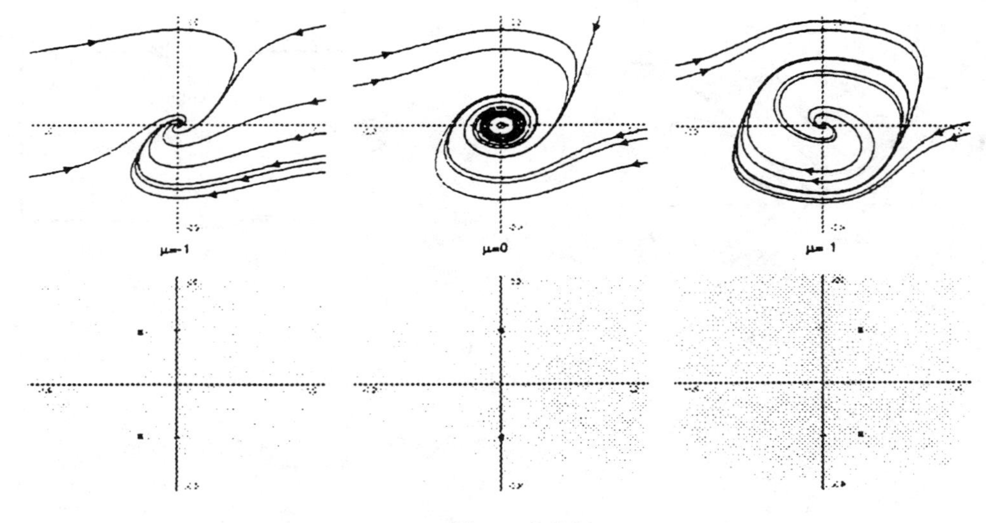

Figure 3.5.2

system

$$\begin{aligned}\frac{dx}{dt} &= x - y - x(x^2 + y^2)\\ \frac{dy}{dt} &= y + x - y(x^2 + y^2)\end{aligned}$$

The only critical point is $(0, 0)$, and it is a spiral source. If an annulus is constructed having an inner radius < 1 and an outer radius > 1, then by examining the vector field on the boundary of the annulus, it's not hard to see that solutions near $(0, 0)$ and solutions outside the annulus eventually enter the annulus and do not leave. Thus it contains a compact limit set, and so a limit cycle (see Figure 3.5.3).

One can also show that, as in our example, any closed solution curve must enclose an equilibrium point, so long as it encloses an open set on which the vector field, F, is defined everywhere. But first we need a definition.

Definition: Suppose γ is a continuous, simple, closed curve in the plane, and F is a continuous vector field defined on γ. If we observe the vector

$$\frac{F(\gamma(t))}{\|F(\gamma(t))\|}$$

as t varies so as to take us once around γ, then this vector will travel around the unit circle an integral number of times, n. We put $I(F, \gamma) = \pm n$, the index of F along γ, taking the plus sign if F is winding in the same direction as γ, and the minus sign if it is winding in the opposite direction.

It can then be shown that

Lemma. *If γ is a C^1 simple closed curve and F is a continuous, non-zero vector field defined on the closed region bounded by γ, then $I(F, \gamma) = 0$.*

Theorem (Umlaufsatz). *If γ is a C^1 simple closed curve with a non-zero tangent vector field, F, then $I(F, \gamma) = 1$.*

For proofs, see [P].

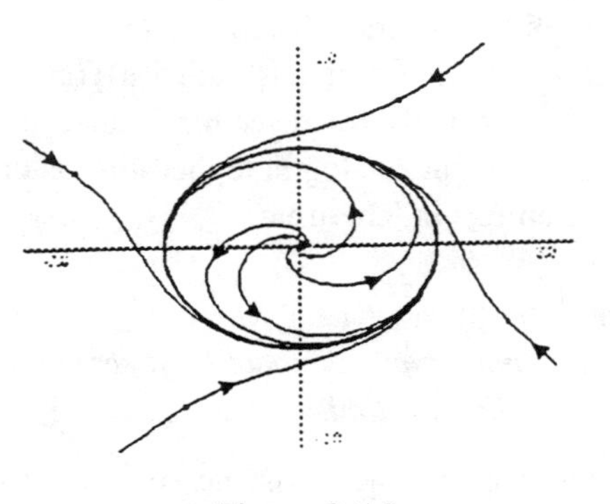

Figure 3.5.3

Corollary. *If*

$$\frac{dX}{dt} = F(X)$$

where $F : U^{open} \subset R^2 \to R^2$ is C^1 and has a periodic orbit, $\gamma \subset U$, then γ encloses at least one equilibrium point.

Thus, when one comes across periodic solution curves in the phase plane, it is natural to inquire after the equilibrium points contained therein, and if we do, we will find that there is an interesting relationship among them. If there are finitely many of them, say $\{x_1, x_2, \ldots, x_k\}$, one can show

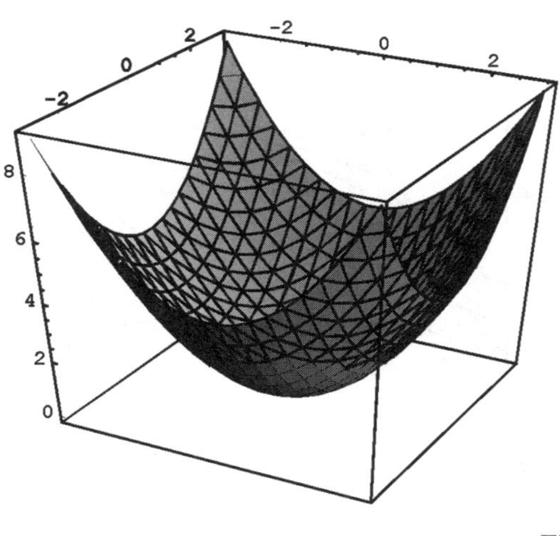

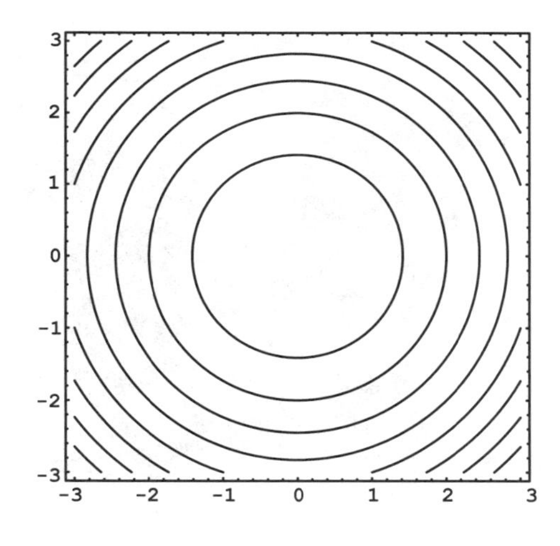

Figure 4.1.1

Theorem. $I(F,\gamma) = \sum I(F, x_i)$.

Again, see [P] for a proof.

Since $I(F,\gamma) = 1$, this places restrictions on the indices of the x_i. For example, one cannot have a periodic orbit which encloses only a saddle equilibrium, or which encloses two sinks. In the first case, $\sum I(F, x_i) = -1$, and in the second, $\sum I(F, x_i) = 2$.

Finally, we note that limit cycles of systems of differential equations in the plane are the subject of the unsolved 16th problem of Hilbert: Given the set of polynomial second order differential equations of degree n, put an upper bound on the number of limit cycles. See [V] for details and partial results.

4. Special Systems

4.1 Hamiltonian Systems

Recall the spring-mass system:

$$\begin{aligned}\frac{dx}{dt} &= y \\ \frac{dy}{dt} &= -x\end{aligned}$$

Put

$$H(x,y) = \frac{1}{2}(x^2+y^2).$$

Notice that

$$\begin{aligned}\frac{d}{dt}(H(x(t),y(t))) &= \frac{\partial H}{\partial x}\frac{dx}{dt} + \frac{\partial H}{\partial y}\frac{dy}{dt} \\ &= x\frac{dx}{dt} + y\frac{dy}{dt} \\ &= xy - yx \\ &= 0.\end{aligned}$$

Thus, $H(x,y)$ is **constant** as a function of time. This means that the level curves of H correspond to solution curves of the differential equation. Physically, H measures the total energy of the system (kinetic + potential). Figure 4.1.1 is a graph of H and its level curves.

The surface is sometimes called the energy surface for the system. Along any solution curve, the energy of the system is conserved.

Notice that our system could have been written as

$$\begin{aligned}\frac{dx}{dt} &= \frac{\partial H}{\partial y} \\ \frac{dy}{dt} &= -\frac{\partial H}{\partial x}.\end{aligned}$$

Definition: If $H : R^2 \to R$ is a function with continuous first partial derivatives such that

$$\begin{aligned}\frac{dx}{dt} &= \frac{\partial H}{\partial y} \\ \frac{dy}{dt} &= -\frac{\partial H}{\partial x}\end{aligned}$$

then the system of DE's is called a **Hamiltonian System**. H is called a **Hamiltonian** function for the system.

Theorem. *The level curves of the Hamiltonian function for a Hamiltonian system correspond to solutions to the system.*

Proof.

$$\begin{aligned}\frac{d}{dt}H(x,y) &= \frac{\partial H}{\partial x}\frac{dx}{dt} + \frac{\partial H}{\partial y}\frac{dy}{dt} \\ &= \frac{\partial H}{\partial x}\frac{\partial H}{\partial y} - \frac{\partial H}{\partial y}\frac{\partial H}{\partial x} \\ &= 0.\end{aligned}$$

Thus, H is constant along solution curves. ■

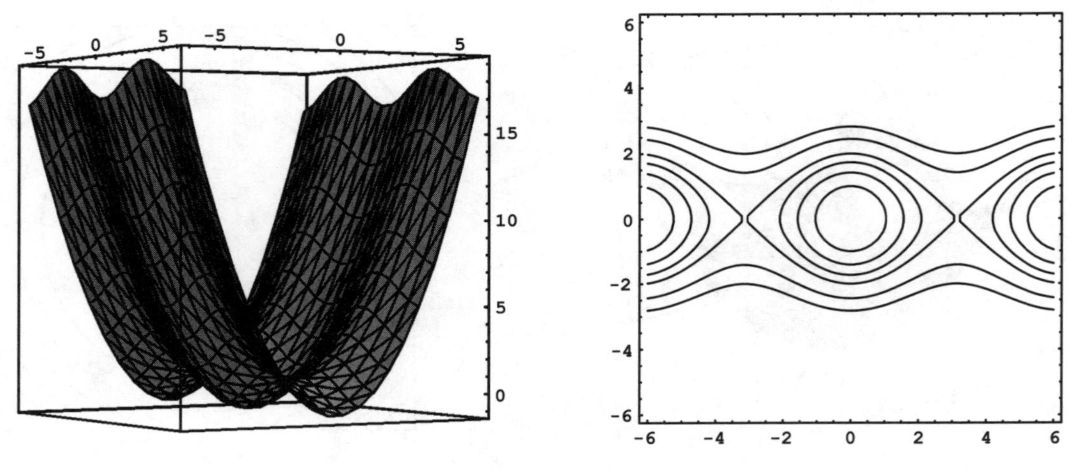

Figure 4.1.2

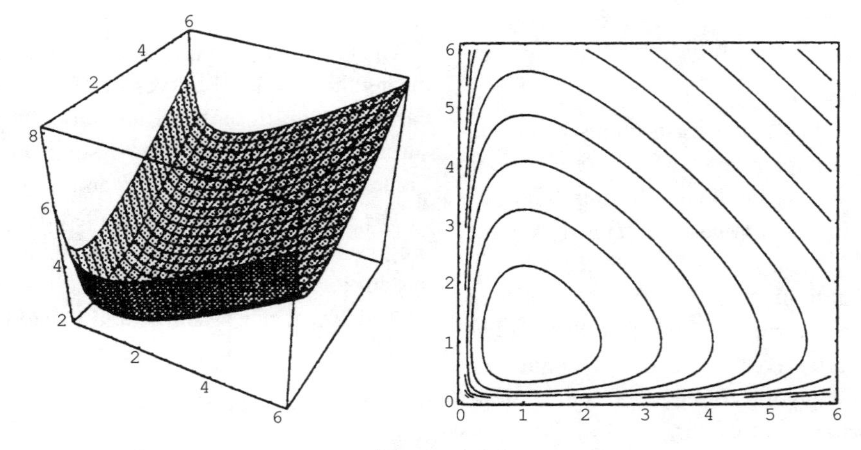

Figure 4.1.3

Consider the frictionless pendulum:

$$\begin{aligned}\frac{dx}{dt} &= y\\ \frac{dy}{dt} &= -sin(x)\end{aligned}$$

The Hamiltonian is

$$H(x,y) = \frac{1}{2}y^2 - cos(x).$$

It is worth noting (and left to the reader as an exercise) that Hamiltonian systems are area-preserving in the sense that, if one starts with a region of the phase space with a smooth boundary, say R_0, and puts $R(t) = \{\phi_{x_0}(t) \mid x_0 \in R_0\}$, where ϕ_{x_0} is a solution curve through x_0, then the area of $R(t)$ is the same as the area of R_0. From this one can see that Hamiltonian systems will have no sources or sinks.

The Hamiltonian is a special case of what is called a **First Integral** or **Constant of Motion**.

Definition: $E : R^n \to R$ is called a **First Integral** or **Constant of Motion** of a system of differential equations $\frac{dX}{dt} = F(X)$ iff E is constant on solutions to the system. That is, for any solution $X(t)$ to the system, there is a constant, C, such that $E(X(t)) = C$.

Theorem. *If the gradient of E (that is, E') is perpendicular to $F(X)$ along solution curves to the system $\frac{dX}{dt} = F(X)$, then E is a Constant of Motion for the system.*

Proof.

$$\begin{aligned}\frac{d}{dt}E(X(t)) &= E'(X(t)) \cdot X'(t)\\ &= E'(X(t)) \cdot F(X)\\ &= 0\end{aligned}$$

Thus, $E(X(t))$ is constant. ■

Consider the system

$$\frac{dx}{dt} = ax - bxy$$

$$\frac{dy}{dt} = bxy - cy$$

These are, of course, the Lotka-Volterra Predator-Prey Equations [BDH]. The reader can check that

$$E(x_1, x_2) = bx - cln(x) + by - aln(y)$$

is a Constant of Motion for the system.

4.2 Dissipative Systems

If we again start with the Hamiltonian spring-mass system

$$\begin{aligned}\frac{dx}{dt} &= y \\ \frac{dy}{dt} &= -x\end{aligned}$$

but change its vector field by adding a damping term, we can get the system

$$\begin{aligned}\frac{dx}{dt} &= y \\ \frac{dy}{dt} &= -x - 0.1y\end{aligned}$$

As would be expected, the system now loses energy and spirals in to an equilibrium (see Figure 4.2.1).

In fact, if we calculate the total energy along solution curves, as measured by the Hamiltonian, we can see that, except at the equilibrium point $(0, 0)$, it is decreasing:

$$\begin{aligned}\frac{dH(x(t), y(t))}{dt} &= \frac{\partial H}{\partial x}\frac{dx}{dt} + \frac{\partial H}{\partial y}\frac{dy}{dt} \\ &= -0.1y^2 \\ &\leq 0\end{aligned}$$

This is an example of a **Dissipative System**:

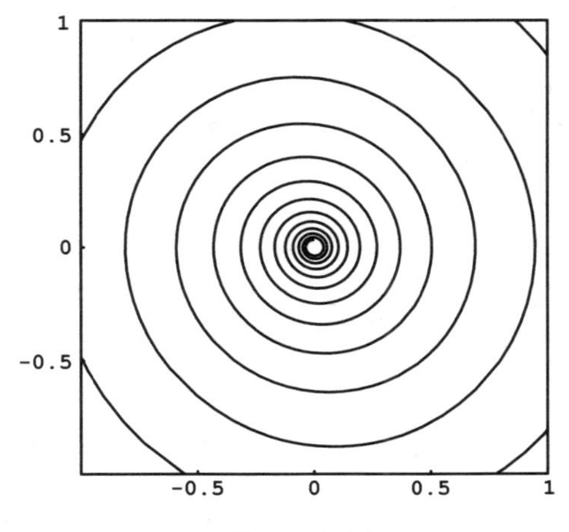

Figure 4.2.1

Definition: A system $\frac{dX}{dt} = F(X)$ is called **dissipative** if there exists a bounded subset of the phase space, B, such that for any point X in the phase space, there is a time t_0 (possibly dependent on X and B) such that the solution through X is in B for all $t \geq t_0$.

One method of establishing that a system is dissipative is to construct, as we did in our example, a **Liapunov function**:

Definition: Let V be a real-valued, C^1 function defined on the phase space for the system $\frac{dX}{dt} = F(X)$. If for every solution $X(t)$ which is not an equilibrium, we have $V(X(t)) > 0$ and $\frac{dV(X(t))}{dt} \leq 0$ for all t, while V vanishes at equilibria, then V is called a **Liapunov Function** for the system. If the inequality is strict, V is called a **strict Liapunov function** for the system.

Notice that in the case of a strict Liapunov function, the solution curves are required to cross the level curves of V, going from higher to lower values of V. This precludes any periodic solutions.

Consider a damped hard spring

$$\begin{aligned}\frac{dx}{dt} &= y \\ \frac{dy}{dt} &= -(x + x^3) - y\end{aligned}$$

We modify the energy function for this system to get

$$V(x, y) = \frac{1}{2}y^2 + \frac{1}{2}(x^2 + \frac{x^4}{2}) + \frac{1}{2}(xy + \frac{x^2}{2})$$

A calculation shows that

$$\frac{dV}{dt} = -\frac{1}{2}(x^2 + x^4 + y^2)$$

so this is a strict Liapunov function for the system. Had we not modified the energy function, the solution curves would have been tangent to the level curves of the Liapunov function along the x-axis. Figure 4.2.2 shows the level curves for the Liapunov function, along with a typical solution curve which cuts across the level curves as it spirals inward toward the equilibrium.

Liapunov functions are of further use in establishing the stability of equilibrium points. One can show that the existence of a Liapunov function implies the existence of a stable equilibrium point (nearby solutions stay nearby for all time) and the existence of a strict Liapunov function implies the existence of a strictly stable equilibrium point (nearby solutions tend asymptotically toward the equilibrium).

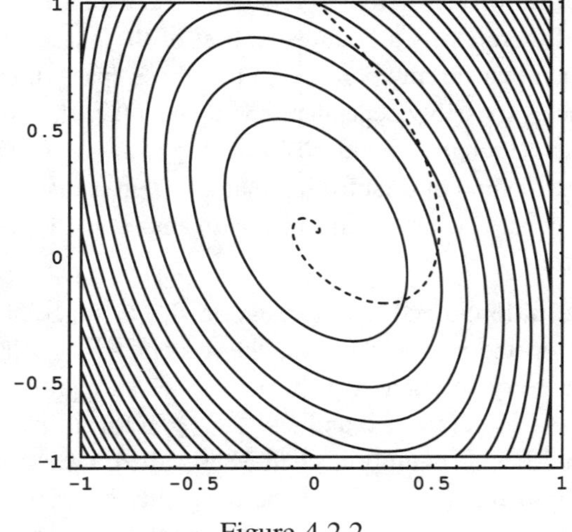

Figure 4.2.2

4.3 Gradient Systems

If we again start with the Hamiltonian spring-mass system

$$\frac{dx}{dt} = y$$

$$\frac{dy}{dt} = -x$$

but replace its vector field by one which is orthogonal to it, we can get the system

$$\frac{dx}{dt} = x$$

$$\frac{dy}{dt} = y$$

Notice that if we define P by

$$P(x, y) = -\frac{1}{2}(x^2 + y^2)$$

then

$$\frac{dX}{dt} = -\nabla P(x, y).$$

(It is conventional to use the minus sign.) Thus, the vector field described by the differential equation will point in the direction of maximum increase for the function P, and is perpendicular to the level curves of P. The critical point, $(0, 0)$, of P is a relative minimum, and corresponds to the equilibrium point of the system, which is a source.

We make the following definition, which generalizes our earlier one for the potential function:

Definition: Let $P : R^n \to R$ be a function such that $\frac{dX}{dt} = -\nabla P$. This system of differential equations is called a **gradient system**. The function P is called a **potential function** for the system.

The introduction of the minus sign will cause the solution curves to become paths of **steepest descent** for the surface that is the graph of $-P$. They will still be perpendicular to the level curves of $-P$.

Consider

$$\begin{aligned} \frac{dx}{dt} &= 3x^2 - y \\ \frac{dy}{dt} &= -x. \end{aligned}$$

If we put

$$P(x, y) = -x^3 + xy,$$

then

$$\nabla P(x, y) = \begin{bmatrix} -3x^2 + y \\ x \end{bmatrix},$$

which means that this is a gradient system with P as the potential function. The solution curves are curves of steepest descent for the surface that is the graph of P. They are perpendicular to the level curves of P. The critical points of P correspond to the equilibrium points of the system. In this case, the critical points are obtained by solving

$$\begin{aligned} (0, 0) &= \nabla P(x, y) \\ &= (-3x^2 + y, x) \end{aligned}$$

so $x = 0$ and $y = 0$.

We could classify the equilibrium point $(0, 0)$ by the usual linearization method. We could also, however, use the following theorem from multivariable calculus, recast in terms of potential functions.

Theorem. *Suppose (x_0, y_0) is an equilibrium point for the gradient system $\frac{dX}{dt} = -\nabla P$. Let $H_P(x, y)$ be the Hessian matrix for $-P$ at (x, y). Then (x_0, y_0) is a source if $det H_P(x_0, y_0) > 0$ and $tr H_P(x_0, y_0) > 0$. Furthermore (x_0, y_0) is a sink if $det H_P(x_0, y_0) > 0$ and $tr H_P(x_0, y_0) < 0$. We also have the result that (x_0, y_0) is a saddle fixed point if $det H_P(x_0, y_0) < 0$.*

In our example, we need to consider the Hessian

$$\begin{aligned} H_P(x, y) &= \begin{bmatrix} -P_{xx}(x, y) - P_{xy}(x, y) \\ -P_{yx}(x, y) - P_{yy}(x, y) \end{bmatrix} \\ &= \begin{bmatrix} 6x & -1 \\ -1 & 0 \end{bmatrix} \end{aligned}$$

At $(0, 0)$, we get $\det H_P(0, 0) = -1$ so $(0, 0)$ is a saddle fixed point.

Here are the level curves of P; the solution curves are orthogonal to them. A zoom in on the graph of P near

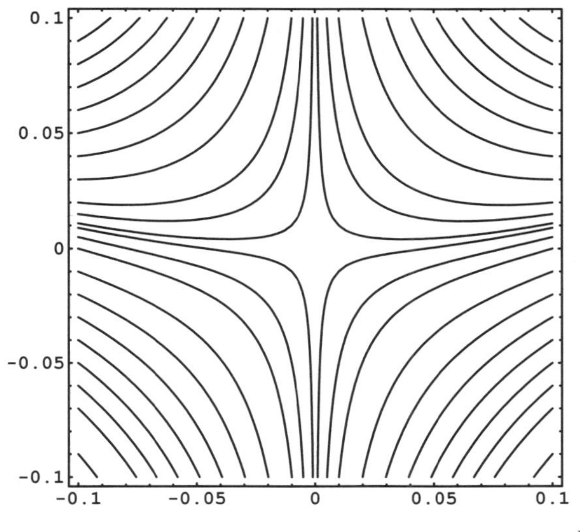

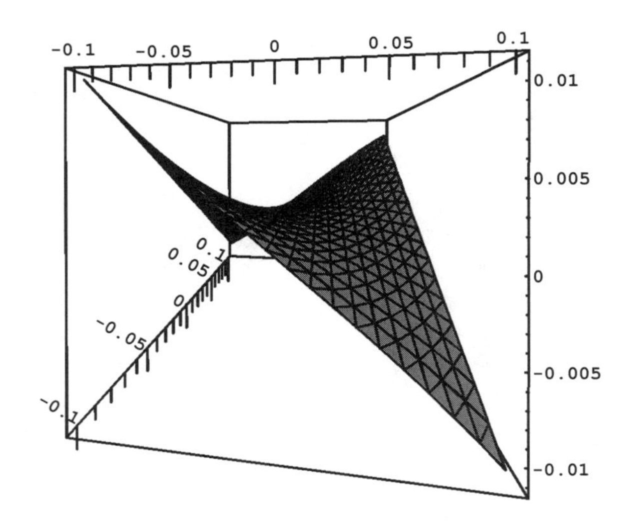

Figure 4.3.1

the origin, shows that it is indeed a saddle near there (see Figure 4.3.1).

Notice that since the Hessian matrix is always symmetric, its eigenvalues are real, and so a gradient system will not have any centers or spiral equilibria.

Further, using the chain rule, for a gradient system we have

$$\frac{dP(x(t), y(t))}{dt} = -\|\nabla P\|^2$$

which means that P is a strict Liapunov function for the system and thus no gradient systems have periodic solutions.

References

[BC] Borrelli, R., and Coleman, C., *Differential Equations: A Modeling Perspective*, John Wiley, 1996.

[BD] Boyce, W., and Di Prima, R., *Elementary Differential Equations*, John Wiley, 1992.

[BDH] Blanchard, P., Devaney, R., and Hall, G., *Differential Equations*, Brooks/Cole, 1996.

[HK] Hale, J., and Kocak, H., *Dynamics and Bifurcations*, Springer-Verlag, 1991.

[HS] Hirsch, M., and Smale, S., *Differential Equations, Dynamical Systems, and Linear Algebra*, Academic Press, 1974.

[HW] Hubbard, J., and West, B., *Differential Equations: A Dynamical Systems Approach*, Springer-Verlag, 1991.

[P] Plante, J., *Introduction to Qualitative Theory of Differential Equations*, Carolina Lecture Series, University of North Carolina, Chapel Hill, 1976.

[V] Verhulst, F., *Nonlinear Differential Equations and Dynamical Systems*, Springer-Verlag, 1990.

Differential Equations on the Internet

Kevin D. Cooper* and Thomas LoFaro*
Washington State University

Abstract

The rapid ascent of the Internet to common use has broad significance for education in general, and for instruction in differential equations in particular. The chief innovation offered by the common use of the Internet is the free and rapid exchange of up-to-the-minute information. This provides powerful library resources for even the smallest academic institution, resources that may be used to develop realistic applications and exercises in a subject that for too long has languished in rote learning and outmoded techniques. Moreover, the network provides didactic software and insight not previously available to institutions that were unable to make a commitment to creative instruction.

1. Introduction

A number of events have shaken the firmament of mathematics, and the field of differential equations in particular, in this century. In differential equations, there has been a tendency to de-emphasize formal solutions to individual equations, in favor of discovering qualitative information about classes of solutions. This approach has the double advantage of admitting analysis of the behavior of solutions to a much larger group of equations, while simultaneously providing geometric insight into that behavior that frequently provides as much information as the actual computation of an individual solution.

Even as this transformation in attitude to differential equations has taken place, the field of digital computing has advanced immeasurably. Computers capable of computing good approximations to solutions of extremely complicated differential equations now rest on the desks of most undergraduate students. Such machinery, when coupled with the right software, easily delivers the visualizations natural to the qualitative approach.

*This work was partially supported by NSF grant number DUE-9555228.

These topics have been amply discussed in preceding chapters. A third development has come to prominence with the rapidity and brilliance of a comet: the Internet. The Internet is composed physically of tens of thousands of computers connected by a variety of media, all sharing certain communications protocols. However, the Internet is more than physical and logical connections: it is a means by which ideas may be distributed instantly to an enormous audience. Moreover, most institutions now provide at least limited access to the Internet to their students. Thus the poorest student in the most deprived university may have access to an enormous library of the most current information imaginable.

This new access to current information has a double effect on the teaching of differential equations. First, it enables rapid communication among instructors of differential equations, for the exchange of teaching ideas, instructional materials, and software. Second, it provides access for both students and instructors to resources that permit development and solution of realistic problems in differential equations. Data sets and graphs are available that permit conjectures to be formed, models to be developed, and results of predictions to be validated, all in a framework of discovery and innovation.

In Section Two a brief history and description of the Internet appears, together with a discussion of the resources and software in common use. The following Section is subdivided into three parts, and offers examples of the kind of exercises that may be developed using Internet resources. The concluding section discusses Internet sites that offer instructional materials and act as clearinghouses for differential equations instruction.

2. Internet Tools

The topography of the Internet changes so rapidly as to make any discussion of the state-of-the-art invalid the moment it is published. Nonetheless, a basic discussion of the ingredients of the network is required in order to understand its potential.

Computing in the seventies was dominated by a "mainframe" mentality. A single large computer was linked to tens or hundreds of terminals, which could use its computational cycles simultaneously. In the eighties, the Department of Defense funded linkage of a number of these mainframe computers together in a project called the ARPANet. This project required development of networking protocols to facilitate communication among different kinds of computers, and standards for the physical layer of the network. While other networks already existed, or were in development, notably by Digital Equipment Corporation, the ARPANet model quickly became the standard. By the late 1980s, most academics used electronic mail regularly, and were familiar with the notion of logging onto a remote computer from a local one. As the network outgrew its Defense Department origins to encompass a broader spectrum of enterprises, and as it subsumed other competing networks, it became known as the Internet. The name Internet thus refers to a certain physical collection of wires, optical fibers, and computers, which share common communications protocols and addressing schemes.

The revolution in the use of the Internet arose with the idea of a browser program. An Internet browser is a program that runs on a local computer, which is capable of communicating with another computer, which may be anywhere in the world. The browser obtains from that computer a quantity of information, and then displays the information in a useful format. The information transferred may be text, graphics, programs, sound recordings, movie clips, or combinations of these. The beauty of the idea stems from a client-server structure. The client (the browser program) requests information from the remote computer (the server). The server delivers the information as its own usage and network access permits, and then the client program displays it in real time. Thus, the image that is displayed may be refreshed by the local computer, the program that was retrieved from the server is not slowed by a busy network, and the movie clip proceeds at the proper rate. Once retrieved, all of this information may be stored on a local file system for future use.

One of the first Internet browsers was created by the National Center for Supercomputing Applications (NCSA), and was called Mosaic [1]. It was available free of charge by anonymous FTP, and was instrumental in popularizing Internet browsing. After Mosaic came into common use, the Internet, as used by browser programs, was frequently referred to as the World Wide Web (WWW), an appellation which frequently is abbreviated to "the Web". Two of the people who worked on Mosaic went on to extend the work independently, and created a new browser called Netscape Navigator [2]. They later developed a commercial enterprise around this browser. Other browsers have been developed more recently, most notably the Microsoft Internet Explorer [3], and Sun Microsystems' HotJava. As this chapter is written, Netscape is the most popular Internet browser, although the Internet Explorer is closing fast, and special computers designed specifically to run the HotJava browser may revolutionize the corporate desktop.

In the early '90s, Sun Microsystems looked at the growing popularity of Internet browsing, and noticed that it was largely static. In other words, the browser program could download a document from another site, and the document could display images or even movies, but there was little possibility for interaction. Programs could not be run from an Internet browser, nor even easily downloaded from another site in a form that would permit them to be run on a local computer. As Sun began to develop methods of downloading and running programs with a browser, it observed that extant programming languages were not designed with the Internet in mind. While languages such as C++ could easily import classes from other files on their current host computer, they had no easy way of importing files from other computers connected by a network. Worse yet, the proliferation of different proprietary versions of operating systems made it difficult to write a computer code that would compile and run on a variety of platforms. This led Sun Microsystems in 1995 to release a new programming language called Java [4]. Java incorporates many of the desirable features of C++ with a somewhat simplified syntax, and the ability to refer to files on remote computers almost as easily as files on the local machine. Moreover, Java has been assembled with numerous widgets to simplify the construction of applets—small programs that may be run within an Internet browser. The result is that one may download a page over the Internet, and with the page may come a program, which runs on the same machine as the browser program. Netscape has incorporated Java since version 2.0. Other languages designed with the Internet in mind have appeared more recently, notably Netscape's own Javascript, and Microsoft's ActiveX.

The effects of all of these developments on differential equations instruction are twofold. First, a wealth of information is available over the Internet that may be used by students and instructors to develop and validate differential equations models. Instead of copying facts from an exercise in a textbook onto their page, and then following a template example to solve a problem, students may go directly to a source of information, retrieve that which they need, and develop their model from there. Second, while interactive textbooks have been discussed for a long time, previously they have always hinged on the use of software with a limited distribution, whose use might be confined to few platforms. The software might have been expensive and difficult to use. With the ascendancy of the Internet and languages designed for it, interactive textbooks are now only a browser page away. They no longer have to be developed by a single publisher, or by a limited group of authors. Indeed, there are projects existing now with the aim of creating large bodies of differential equations materials by authors from around the world.

There is one issue here that colors all use of the Internet. The number of users on the Internet, the amount of network traffic, the number of Web sites, and the size of the applications, are all growing dramatically faster than the technological infrastructure that supports them. As a result, many ideas that look good on paper prove to be unreasonable in practice, due to slow transfer times and intermittent failures in components of the network. It will be important for some time to come to keep Web applications "lean and mean" in order to make them genuinely useful.

In the next section, we will discuss three approaches to use of the Internet in a differential equations course. These range from approaches in which all the resources a student needs are obtained over the Internet, to other approaches in which the Internet is nothing more (or less) than a library. The last section will discuss some specific sites that provide resources for differential equations on the Internet.

3. Examples

Salmon Migration

Anyone who has used a "real-life" application in a class, with genuine data from modern scientific experiments, knows that students thrive on the exercise. Students of differential equations always want to know that they are learning skills applicable to their future careers. Showing them a modern example using actual data encourages them to take a greater interest in the subject. Moreover, real problems are never so cut-and-dried as textbook problems. Textbook problems in differential equations frequently encourage rote learning and imitation, while solving real problems requires some real understanding.

Formerly, an instructor who wanted to use a real problem in class had either to condense the data and give it to the students in an abridged form, or else send the students to the University library—a library which would have to be outstanding if it were to have the resources required. The material in the library would presumably be dated, perhaps even invalid, due to the time required for publication and distribution. Moreover, unless the instructor provided specific information regarding the whereabouts of useful information, the student would spend a great deal of time finding and sorting sources. Abridged information provided by the instructor suffers from the same drawbacks, except that in this case, it is the instructor that must spend the time locating and sifting the information. Abridging the data carries the added disadvantage that the student no longer has the feeling that she is dealing with the "real world." The fact that the instructor has processed the information renders it less exciting.

In contrast to this, the Internet now makes a large volume of data available to everyone. Some of the data have already been processed, while other sets appear on the Web even as they are measured. Thus, exercises may be developed to model phenomena using data that were obtained the day of the exercise. The search and sifting must still be performed, but modern Web search engines can expedite the process immeasurably.

As an example of this use of the Internet, consider a naive model for the flushing of juvenile salmon and steelhead, called smolts, down the Snake River in Washington. This problem has been in the news in recent years, since the dams on the Snake and Columbia Rivers inhibit the migration of smolts to the sea. In order to get the smolts to the Pacific, various approaches have been tried, including barging them down the river, spilling water from the dams on the river so as to create a more powerful current and reduce the volume of the reservoir, and combinations of these ideas. We are interested in approaches that involve spilling water. Historically, the reservoirs behind the dams are filled in the spring, and the water is used for irrigation, navigation, and hydroelectric power throughout the rest of the year, particularly during the summer. On the other hand, the smolts migrate in late spring and early summer, so that if the Snake is to be a river, then water may not be stored in the customary way. Instead, the outflow at the dam from each reservoir must be as great or greater than the inflow. In actual practice, the reservoirs have sometimes been "drawn down", meaning that much of the water from the reservoirs is flushed out.

Why is it necessary to empty the reservoir? Would it not be sufficient to make the outflow equal to the inflow, so that the reservoir would behave like a wide spot in the river? A very simple compartmental model might be used to illustrate the ideas behind the drawdown.

Consider the Lower Granite Reservoir, near Lewiston, Idaho, formed behind the Lower Granite Dam. It is fed by two rivers: the Snake, and the Clearwater. The Snake is fed, in turn, by the Salmon River, not far from Lewiston. We can obtain both average and current streamflow data for these rivers by using the USGS web pages [5] that provide indices to streamflows in Idaho. Flows for the Clearwater near Spalding, Idaho (near the reservoir) are included. It does not provide a streamflow for the Snake sufficiently near Lewiston to be of use directly, but we can add the Snake flow near the Hell's Canyon Dam to the flow from the Salmon river near Whitebird, Idaho, to arrive at a very rough approximation of the flow of the Snake into the Lower Granite Reservoir. The total flow is denoted I. This varies from month to month, and indeed from day to day, due to the timeliness of the data. The quantity I provides an input to model the volume of water in the reservoir at any time.

Salmon smolts are released near the headwaters of the Salmon and Clearwater Rivers, as well as in other tributaries of the Snake River. In addition, a small population of wild fish still use these rivers to spawn. The smolts swim and are swept downstream, eventually coming to the Lower Granite Reservoir. Let $s(t)$ denote the number of salmon in the reservoir at time t. There is a wealth of information on the Internet concerning salmon migration into the Lower Granite Reservoir [6]. By consulting a fascinating collection of Web pages sponsored by the University of Washington School of Fisheries [7], we may learn that smolts find their way into the reservoir at a more-or-less constant rate, denoted r_s, during the migration period. We assume that their swimming abilities allow them to disperse themselves uniformly throughout the reservoir after entry, providing the usual "well-stirred tank" of a standard compartmental model. The rate at which salmon are swept into the reservoir depends only slightly on the rate of flow of the rivers. It depends more strongly on the rate at which smolts are released. On the other hand, the rate of flow of salmon out of the reservoir depends on the rate of flow of water over the spillway and through the turbines, and on the concentration of salmon in the outflow. Letting the volume of the reservoir be denoted V, the concentration of salmon in the reservoir is $s(t)/V$. Letting the rate of flow out of the reservoir (which the Army Corps of Engineers [8] can control) be denoted F, a differential equation for the flow of salmon smolts out of the Lower Granite Reservoir has the form

$$s' = r_s - \frac{Fs}{V}. \tag{1}$$

We were unable to find information on the Web directly giving values for V. However, plausible estimates might be obtained using pool elevation data [9] together with some geographic information from maps. It might also be derived using a second model:

$$V = I - F, \qquad V(0) = V_0. \tag{2}$$

The inflow and outflow data are available, as already discussed, and an initial condition V_0 might be obtained using a period of extreme drawdown, as occurred in 1993.

The rate at which smolts are flushed downstream increases as the volume of water in the reservoir decreases, which is the reason the reservoir is drawn down. The student will be able to compute various scenarios for the drawdown after solving the equation. This will be easily solved by students of differential equations if r_s, V, and F are constant. More involved exercises may be developed by using either real data, or estimates derived from data, to make those coefficients functions of time. However, the real point of this discussion is that the Internet has made it possible to find a wealth of current realistic information for a differential equations problem. This information does not enhance the mathematical content of the problem significantly, but finding it lends an air of immediacy and verisimilitude to the problem that a student could never obtain by reading the problem out of a textbook. Experience shows that this enhances a student's interest in the subject matter markedly.

While the advantages of such realistic exercises should be apparent to instructors, so should the difficulties involved. Formulating such a problem is a time consuming process. Instructions for dealing with real data must be very precise, or else students will become confused and will be unable to do the problem. It is always necessary to make estimates of some of the quantities actually required, since such quantities are frequently difficult to measure, or are not easily available. The very volatility in the information that makes the problem exciting to the student means that such an exercise has a very short life span. Sites referenced may disappear from one academic term to the next. On the other hand, such an exercise, once formulated, may be made available to a large group of instructors immediately, by using the Internet. In other words, the Internet provides a medium for a consortium of differential equations instructors at scattered institutions, or for publishers interested in a subscription service, to formulate and distribute exercises.

Finally, it is worth noting that this approach to the Internet is relatively conservative. The Internet is simply used as a library whose contents are constantly updated. Students must still work problems out by hand, or by using some resource independent of the browsing program, and presumably write their results on paper for submission to the instructor. In short, the Internet provides only textual information, but no software or interactive elements. In the next section, we will describe the use of the Internet to obtain and run software for differential equations.

Chemical Oscillations

The Internet may provide more for students of differential equations than a current research library. In particular, most browsers can launch other programs that reside on the hard disk of the local computer. Netscape calls these programs "plugins" The idea is simple: the student associates a certain file type with a certain viewing program on her local computer. Whenever a file of that type is encountered by the Internet browser, then the browser starts the associated program on the local computer, and then loads the file as input to the program. Thus the student ends up with two windows on her screen—one showing a web page, and the other showing the "downloaded" file using the local program. The student may interact with the program that runs locally, possibly following a text template from the browser window.

As an example of this approach, consider the page "Chemical Oscillations" that may be found as part of the IDEA project [10]. This page gives a historical background of the discovery of chemical reactions that oscillate, and then gives several "chemical recipes" for creating oscillating chemical cocktails. Together with this there are differential equations describing the reactions. The page was written by a chemist, and discusses reactions that may actually be reproduced. For example, the "Brusselator" reaction is given, and leads to a pair of differential equations describing the amounts of transient compounds formed as it proceeds.

The periodicity of the system (for certain parameter values) is reflected by the existence of a limit cycle in the model. The limit cycle is depicted in a diagram on the Web page. By clicking the diagram with a mouse, a student receives an input file for the software package MathCad. MathCad is a program for evaluating mathematical expressions, both numerically and analytically, and plotting the results. One might think of it as a free-form, mathematical spreadsheet. It must be purchased and installed on the computer separately from the browser and other Web software. If the browser the student is using is installed properly, then MathCad will start automatically when the data file arrives, will read the data, and display it independently of the browser program. Thus the student will have two windows on his screen: one containing notes and exercises concerning the Brusselator, and the other a MathCad window in which equations and parameters may be manipulated, and trajectories plotted. The student reads the exercise from the computer, and does his work on the computer. This is a more thoroughgoing approach to the use of computers in instruction than that discussed in the previous section. While the computer may not be used exclusively, it is the primary tool for the study of the system in question.

There are some projects that distribute software, in addition to the input files for that software. Indeed, the IDEA project [11] distributes an application that computes and displays numerical solutions to differential equations in two and three dimensions, called DynaSys. Another example of a site that distributes programs is the Mathwright Library [12], a project supported by the NSF, which distributes two packages of software. The first is a program to read certain files containing a combination of mathematics and animated graphics, while the second is an "authoring tool" used to construct those files. To use the Mathwright Library, a student must first download the reader software provided, and execute a few commands to install it on her local computer. After it is successfully installed, she should configure her browsing program to associate the input files with the reader program. Finally, she may visit the Mathwright electronic library, clicking on links to the Mathwright "books". These are self-contained modules that use graphics to illustrate the features of the systems discussed. She may modify equations and experiment with initial conditions, or interact with the program in other ways.

One specific kind of plugin for a browser is a movie viewer, such as an "mpeg player" These programs permit one to download a movie clip over the network, and then play it on the local machine. This permits animations to be recorded and played back at remote sites. For example, one might visit the Institute of Applied Mathematics at the University of British Columbia, and view a short computer-generated movie clip depicting stabilization of a multibody system [13]. Such plugins may provide motion and sound to make a page more useful and interesting.

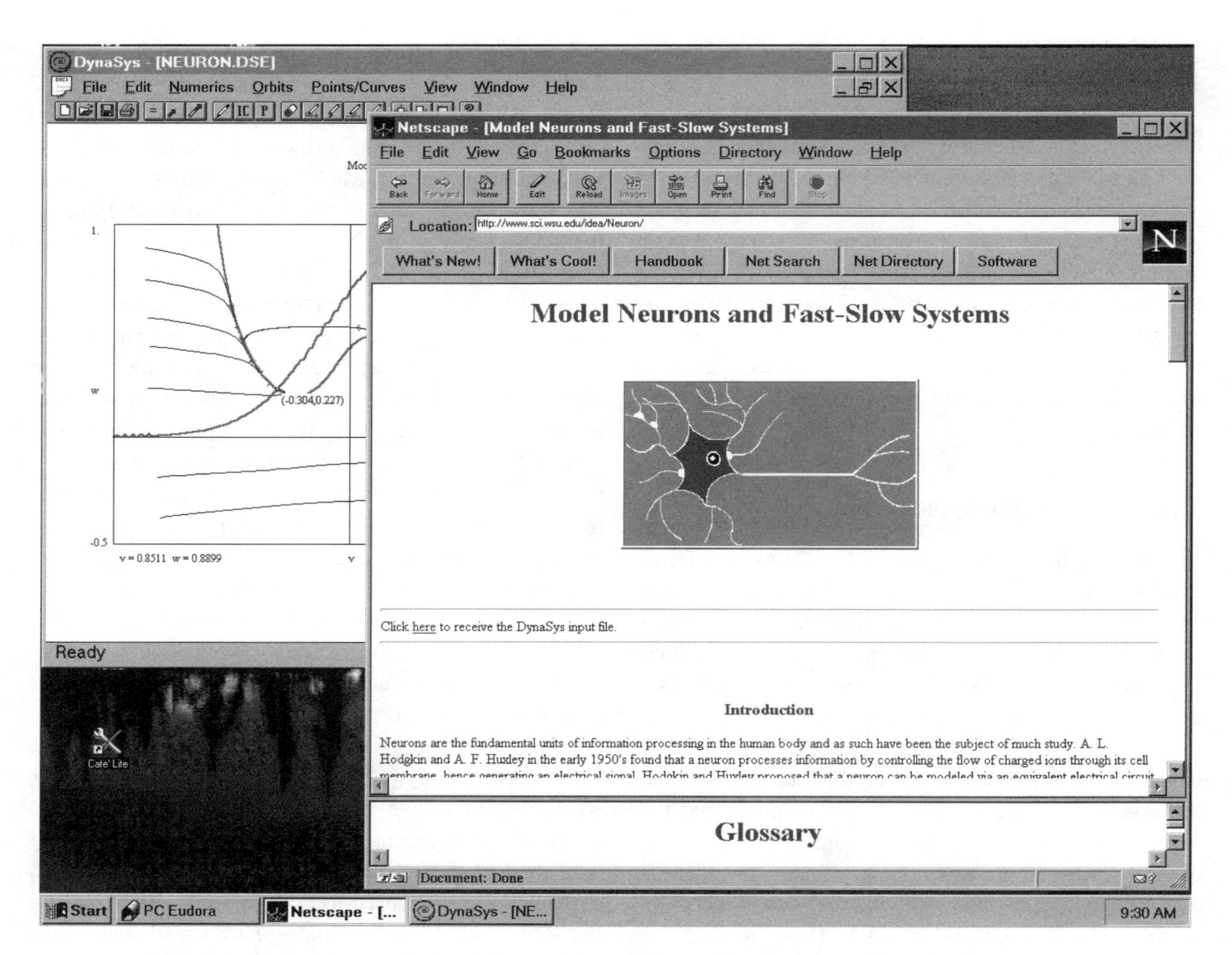

Figure 1. Screen dump of a typical scene in working with a browser and a plugin program.

The advantage to using plugin programs is that they can make full use of the capabilities of a local computer. Processing is as fast as the hardware of the local computer permits, and results are easily printed or saved. One has all the power of an application permanently, without having to wait for it to come across the network each time. In principle, the plugin idea may be used with any software package, including Maple, Mathematica, MathCad, or MATLAB. Indeed, there are a number of sites [14] that use these very packages for a number of topics, not restricted to differential equations. Many of these resources are listed in the Math Archives site [15]. The number of plugins is growing rapidly, so that one may find such a program for almost any purpose desired.

On the other hand, there are a number of disadvantages to the plugin approach. The first problem stems from the proliferation of plugin programs. There are simply too many plugin programs vying for your disk space. Each program and all of its ancillary files must be stored on a hard disk on the local computer, hence the number of plugin programs installed must be limited. To avoid running out of disk space for more important applications, one must choose carefully from among the plugin programs available. In other words, one may be forced to do without the functionality of a plugin program that would be used only occasionally. Moreover, Internet sites that rely on proprietary applications, such as *Maple* or *Mathematica,* require the user to purchase those expensive software packages before using the sites. Such purchases might be prohibitive for many users, and particularly for students working on homework from a home computer.

As an alternative to storing a plugin program permanently, one might choose to install it as needed. Unfortunately, doing so involves the extra steps and awkwardness in installing the required software. While sites that distribute such software try to make the process easy for the user, they nonetheless usually involve some commands in a shell, or at least outside the browser. Typically, one clicks a link in the browser that causes a large file to be sent to the local machine, and the local computer stores it on the hard drive. This process may take from one to thirty minutes, depending on the size of the file and the traffic on the network. Next, one goes to a shell and executes a command to unpack the software, and then one must run a setup program. The setup program may take some time to execute, and may require the answers to questions about the system. When finished with the program, one wants to delete it, but doing so leaves some unneeded settings in configuration files—it is difficult to be entirely rid of the unwanted program.

A third problem regards security. Any time a user downloads a program from a remote site, he takes the risk of loading a virus on his computer. There are no safeguards inherent in the system. One has to use a recent version of a virus scanning program on the local computer, and hope for the best. Worse yet, it is conceivable that one could download a data file designed to tie up a system. Thus, even in an innocuous piece of software containing no viruses, one might load a particular file off the Internet that would cause difficulties of some kind for a machine running that software. The rule for the use of plugin programs is for the student to be sure that he never downloads any file associated with the plugin that does not come from an extremely trustworthy site.

The last problem of plugin programs is that they are compiled, and therefore are platform-dependent. In other words, the student must have the right kind of computer in order to run the plugin. Thus, the authors of plugin software must write versions for several different operating systems, or else risk the danger that the usefulness of their programs will be limited.

Thus far it is difficult to find Web sites that integrate the plugin approach to Internet instruction with the electronic library approach, but the possibilities are obvious. To combine the two would bring the library and the solution tools to the same screen. We look forward to seeing a Web page containing links to a number of interesting and informative sites, with one link that launches a local phase plane program that may be used to tabulate and plot the results from all of the data received.

Bungee Jumping

There is a third approach to the use of the Internet in instruction, which we will call the integrated approach. In this approach, software that is built-in to a browser program is used to generate interactive applications. In previous examples, the browser program did nothing other than retrieve data files from remote computers, and either display them, or save them on the local computer. In the integrated approach, the browser program may retrieve and run programs as well as data from remote machines, and may send data to those machines as well. There are several tools integrated with the most popular browsing programs. The most popular is a product of Sun Microsystems, called Java.

Java is a programming language, in design rather like the C++ language. Unlike C++, Java is not a fully compiled language. Instead, files of code are compiled into a device-independent form that permits them to be sent across the Internet. Upon arrival at another machine, an interpreter on that computer runs the programs. In this way, the problem of platform dependency is largely eliminated. The most popular browsing programs incorporate the Java interpreter, so that if one uses Netscape or the Microsoft Internet Explorer one may run Java code without further work. In particular, it is not necessary to install any programs on the hard disk. Java programs that run on browsers are called "applets"—emphasizing that they should be small applications. There is no limitation on the size of an applet—indeed, some are quite large and sophisticated, but current bandwidth constraints of the Internet dictate that Java programs should be kept small.

There are other languages that run locally. ActiveX is Microsoft's answer to Java. Like Java programs, ActiveX applications may be sent across the Internet to run on a local computer. On the other hand, ActiveX currently runs only on computers that use Microsoft operating systems (though that may change soon). As this is written, Netscape Navigator does not incorporate ActiveX, but the Microsoft Internet Explorer incorporates both ActiveX and Java. A third choice of language in the integrated approach is called Javascript. Javascript is not actually related to Java, and is somewhat less powerful, but it is supported by the Netscape Navigator.

All of these languages represent implementations of "client-side" processing. The idea behind this nomenclature is that in any Internet application, there are two computers involved. The server simply responds to requests for information, formatting it for transfer, and sending it to the requesting computer. The client is responsible for maintaining the windows on the screen, and presenting the information. All of the languages discussed above are implemented so that the information sent by the server may include a program, which then loads and runs automatically on the client computer. Obviously, there is another approach—the program might run on the server computer, which would then send results to the client. This is known as "server-side" processing.

Server-side processing is implemented using the mainstay of interactive computing on the World Wide Web: Common Gateway Interface (CGI) programming. CGI differs from the programming languages discussed above in that it only provides an interface for sending input to a server computer. It generates a data file on the server, which processes the file using instructions written in a standard language such as C or perl. Thus, before Java appeared on the scene, it was still possible to fill out online forms, which could act as input for a program on a server, which would then send the results to the browser on which the form was filled out. CGI remains the principal means of connection for server-side processing.

As an example of the integrated approach to mathematics instruction on the Web, consider a Web page modelling the motion of a bungee jumper [16]. The student might be told by her instructor to do parts 1, 2 and 4 of the computer lab assignment found at the URL given in the reference. She then goes to any computer that runs a browsing program with Java support, and starts the assignment. She finds text (including equations) discussing the model, and problems that demand analysis of the equations of the model. Java applets allow her to draw phase portraits interactively, changing values of parameters as needed. It is not be necessary for her to download any software to her local computer, or do any other technical work beyond learning the browser and the simple Java interface. The Java phase portrait programs themselves are easy to use: she can click any point in the phase plane to draw an orbit using that point as an initial condition. She may select a parameter whose value is to be changed, and change the value by clicking a small button on the applet. The Web page includes words that are linked to glossary entries, and links to more general sites concerning bungee jumping. Thus, the student has all the tools she needs to do the assignment right on her browser program (apart from pencil and paper).

As one might expect, there are a number of disadvantages to this approach as well. The foremost concerns security. Java has been carefully designed so that it is unable to perform read or write operations on the local computer, but security flaws have been found on occasion that allow these restrictions to be circumvented. Whenever such a flaw is found, Sun Microsystems patches it, but nonetheless one must be aware that Java security is not perfect. Javascript suffers from the same kinds of problems. ActiveX lacks many of the security features of Java, and hence is more powerful, but also more dangerous. Our feeling is that Java is safe provided that one visits only reputable Web sites. When "surfing the Internet" i.e., visiting sites whose reliability is unknown, it is safer to turn off the Java capability of the browser program.

A second problem is that the results of a Java program cannot be printed locally. This is due largely to the security features of Java, which do not allow it to read or write to the hard drive of any machine but the one that generated it. Thus, Java applets may be used for screen interaction, but not for anything permanent. ActiveX does not share this problem, but as a result has increased security problems, as discussed above.

Another problem stems from the nature of this kind of programming—the Java applet must be loaded from the remote computer each time a page is visited. This increases the time required to gain access to the Web page, and limits the reasonable size of the applet. A sophisticated applet could easily require so much time to load across the Internet that it would discourage visitors to the site where it is found. Until there is greater bandwidth on the Internet, Java applets must be kept small.

On the other hand, each applet may be customized for the Web page with which it is associated, so that the student has precisely the tools she needs, when she needs them. They maximize simplicity for the user, and Java probably provides more security than is available with most plugin programs. They also maximize simplicity for the programmer, since only one version of the code needs to be written, instead of the many versions required for fully compiled code on proprietary operating systems. This speeds the delivery of software to the desktop, and presumably increases its reliability.

Currently, "network computers" are being offered for sale by several vendors. These machines are designed with the idea that nothing, except a rudimentary operating system, needs to be stored locally. Instead, software is downloaded from any server to which the user has access, and runs locally. Thus the network computer has sufficient memory to run Java applets, but contains no hard drive. Such ideas have been tried before, the most notable example being the "X-terminal" however, the prices on such platforms have always been too high, and their use has been too restricted. The network computer sells for considerably less than an X-terminal, and has access to most of the Internet. These factors may make it the platform of the future for student computing laboratories, and hence for computational differential equations instruction.

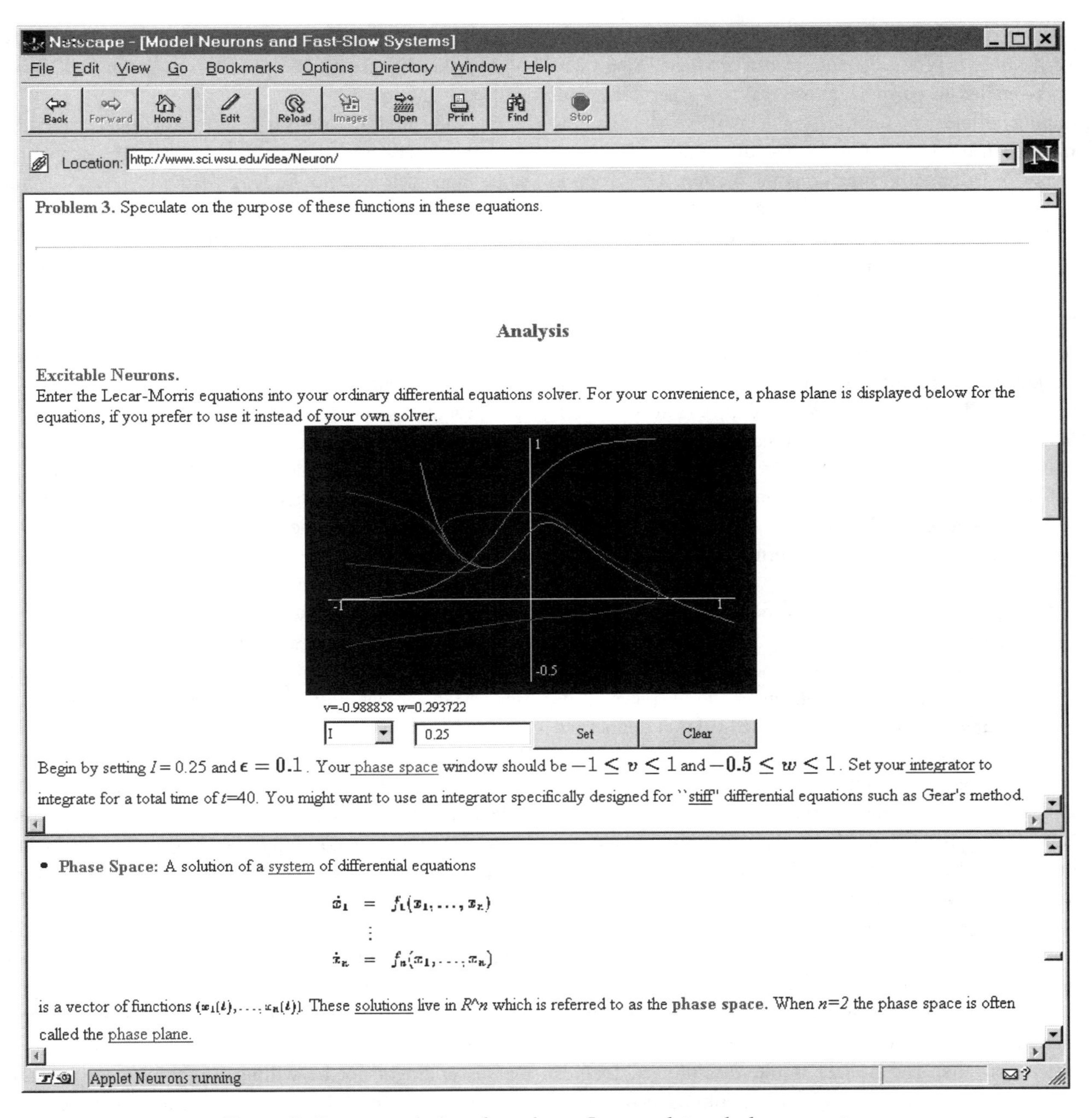

Figure 2. Browser window featuring a Java applet and glossary entry.

4. Differential Equations Web Sites

The amount of useful information available on the Web for a course in differential equations does not yet compare with the amount available for Calculus, but it is increasing steadily. As Calculus reform becomes established, faculty, funding agencies, and publishers are turning their attention to the Differential Equations course as the next frontier. Newcomers to this field may find several reputable sources already established.

One of the oldest groups working on modern approaches to Differential Equations is the Consortium for Ordinary Differential Equations Experiments (C*ODE*E) [17]. This NSF sponsored group, led by members of the faculty at Harvey Mudd College, encompasses representatives from Cornell University, Rensselaer Polytechnic Institute, St. Olaf College, Stetson University, Washington State University, and West Valley

College. The C*ODE*E project has evolved over the years, and has produced a number of products. A newsletter is mailed to participants several times per year, containing articles on topics from actual experiments that have been used in classrooms, to discussions of current issues in Differential Equations Instruction. The group is currently working on a package called ODE Architect, a CD of experiments, together with software for their solution and for further investigations.

The Boston University Differential Equations Project [18] is a second well-established center for instruction in Differential Equations. The originators of this project, leaders both in research and instruction in dynamical systems, have instituted a thoroughgoing reform in the syllabus of their courses. The project is supported by the NSF, and is linked to another project that brings some of the ideas of dynamical systems to high school students. Both the C*ODE*E project and the Boston University Project have resulted in reform texts for Differential Equations.

While they provide some resources using the Internet, neither of the previously mentioned sites focuses strongly on delivery of instructional materials over the Web. Brooks-Cole Publishing, under the auspices of its parent company, ITP, maintains a site [19] that it inherited from PWS Publishing when it was purchased. This site contains some instructional lab activities for Differential Equations, and has links to many more.

One of the newest sites featuring Differential Equations Laboratory exercises on the Web was discussed in the text, and is maintained by Washington State University. An NSF-funded project called IDEA: Internet Differential Equations Activities, supports a site featuring numerous laboratory exercises with Java applets, Differential Equations software that may be downloaded, and a CGI interface to allow construction of Java applets for displaying phase portraits. The project is interdisciplinary, including authors from chemistry and agricultural economics, in addition to mathematicians.

If one is attracted to the plugin approach to Web computing, there is a site called Mathwright [20] that features a number of Differential Equations and Calculus labs, as well as software to generate and display the labs that can be downloaded. This site was also discussed in the text. The Mathwright Player is software that displays workbooks, displaying text, equations, and animations. This site, like many of the others, is funded by the NSF.

Other sites may be found by using one of the standard Web search engines, or by visiting a mathematics clearinghouse site such as Math Archives. There are a number of sites at individual universities, such as the University of Maryland [21]. By the time this is published, there will be many more.

The potential for delivery of instructional materials over the Internet is tremendous, but is only beginning to be tapped. It will not be long before there are several sites that integrate the use of the Internet as a library, as a program-launcher, and as an application interface itself. It is not difficult to imagine the salmon model from Section 3 as a fully interactive lab exercise. The student opens the URL of the site, and is presented with an introductory text, complete with sound and a short movie clip of salmon swimming upstream. When she comes to the first problem, she is provided with a number of URLs at which she may find data from which to compute plausible constants for the model that has been developed. The next problem involves finding an analytical solution to the resulting constant coefficient equation, so she clicks a button to make a Maple window come up, in which she finds and plots an analytic solution. She keeps the browser visible in the background, so as to read instructions and suggestions. Eventually, she comes to a problem involving qualitative analysis of a problem, at which she finds a Java applet built into the page on the browser. She uses it to run some orbits, and to plot the nullclines of the equations for different parameters. She types the text of her answers directly into the browser program, which imbeds them in the page she is reading. When she finishes, she turns in a printout of her Maple workbook, and a copy of the modified HTML page from the browser.

This scenario is not completely feasible using the technology available as this is written. At present, it is not possible for the student to print the results of her work in a Java Applet, and the only reasonable way for her to type her text into the page would involve CGI windows and processing by the server—processing that the sponsor of the server might not be willing to do for a large class. On the other hand, it is inevitable that these features will become available, and that will probably occur soon.

Evidently, one caveat is in order. There is a potential that instructors of differential equations will find themselves concentrating less on the mathematical issues involved than on the technological problems of the delivery medium. It would be easy for instructors to find themselves spending less time talking to students, and more time typing on a computer. Indeed, there are a number of educational institutions that currently are emphasizing the use of the Internet to try to teach more students using the same number of faculty. We believe that such efforts will ultimately fail. There is no substitute for a human voice, and for genuine human interaction, in teaching any subject. Nonetheless, all instructors of differential equations will need to be on their guard to keep sight of the educational goal, and the proper use of the technological tools in reaching it.

No technological development since the invention of the printing press has had the effect on instruction that the advent of the Internet has, and will continue to have. The methods of instruction that these technological developments make possible will, for better or worse, change the face of college instruction. The coming changes would be easier to cope with if they could be anticipated, but it seems that this is impossible. By the time this chapter is published, it will be out of date. New sites will have appeared, some of which will contain new approaches not discussed here. Other sites will have vanished. Indeed, the technology is advancing at such a rate that classical publishing cannot keep up with it. The day has arrived when unprecedented possibilities for rapid communication among instructors, researchers, and students from different institutions allow development and dissemination of much more stimulating and realistic instructional materials for students at every level of the subject.

References

[1] http://www.ncsa.uiuc.edu/SDG/Software/Mosaic

[2] http://home.netscape.com

[3] http://www.microsoft.com/ie/

[4] http://www.sun.com/java/list.html

[5] http://idaho.usgs.gov/public/h20data.html

[6] E.g., see http://www.cgs.washington.edu/dart /dart.html

[7] http://www.cqs.washington.edu/index.html

[8] http://www.nww.usace.army.mil/html/offices/pl/h /wm/rreports.htm

[9] http://www.cgs.washington.edu/dart/pass_rpt.html

[10] http://www.sci.wsu.edu/idea/OscilChem/

[11] http://www.sci.wsu.edu/idea/

[12] http://isaac.engr.washington.edu:80/mathwright /index.html

[13] http://www.iam.ubc.ca/demos/demos-menu.html

[14] E.g. see http://www.thomson.com/pws/pwsftp.html, http://www.cecm.sfu.ca/projects/OMP/

[15] http://archives.math.utk.edu/

[16] http://www.sci.wsu.edu/idea/Bungee/

[17] http://www.math.hmc.edu/codee/main.html

[18] http://math.bu.edu/odes/

[19] http://www.thomson.com/rcenters/diffeq/labs.html

[20] http://isaac.engr.washington.edu:80/mathwright/

[21] http://www.math.umd.edu/schol/ode.html

Data as an Essential Part of a Course in Differential Equations

David O. Lomen
University of Arizona

Introduction

The nature of the first course in ordinary differential equations (ODE's) is rapidly changing. The previous standard course was mostly limited to finding explicit solutions for the few types of ODE's for which such solutions exist. If numerical solutions were covered, the coverage focused on deriving specific algorithms and possibly some error bounds. At many schools this course was aptly called a "plug and chug" course. However, recent advances in hardware and software have provided a means to quickly obtain numerical, graphical, and sometimes closed form, solutions of initial value problems. These advances will continue and within a few years all college and university students will have access to this computer technology. With this situation, which already exists at some schools, the relevance and value of this standard course will be questioned by both faculty and students. This will be especially true at schools where students are accustomed to using graphical and numerical approaches to complement analysis in their calculus courses. These are some of the factors behind the current drive to develop a much more useful and interesting course in ODE's.

The nature of this updated first course in ordinary differential equations will vary from school to school, depending on several factors. Some of these are:

1. Whether all students are required to have their own computer (powerful calculator) or use the school's computers.
2. Whether the school has sufficient space and computers to run the course with a laboratory component. (Also whether the faculty desire such a course.)
3. Whether the school requires all students to be competent in a specific commercial software package such as Mathematica, Maple, or Derive.
4. Whether linear algebra is required as a prerequisite.
5. How much time is devoted to ordinary differential equations in the calculus courses.
6. The requirements of the client disciplines of this course.
7. Whether it is part of an integrated engineering curriculum.

Two of the common threads among the many new types of courses that are currently being developed are a focus on understanding rather than pure manipulation and the use of computers or graphing calculators to enhance graphical and numerical reasoning, as well as to explore solutions of ODE's which have no explicit analytical solution. There is also an increase in the use of mathematical models in several of the endeavors. This paper will focus on the use of data sets to enhance the teaching and learning of ordinary differential equations.

Use of Data Sets

Data sets may be used to develop differential equations, motivate specific topics, suggest proper mathematical models, and check the applicability of such models. These sets may be obtained from other textbooks, research articles, and experiments from science or engineering laboratory courses. Students may also collect their own data using simple equipment (counting number of popcorn kernels popped as a function of time, measuring growth of mold on a piece of bread, measuring the height of water as it drains from a 2 liter plastic soft drink bottle with a small hole in its bottom) or using automatic devices attached to a graphing calculator or personal computer. Manipulation of the data is aided by use of a spreadsheet or many software packages. At our institution, all students have access to a public domain software package TWIDDLE, which allows graphs of functions of up to three parameters to be plotted along with data. It also lets students manipulate data and find a best least square fit with a key stroke. (This, and other public domain packages appropriate for differential equations may be be downloaded from the Web address http://www.math.arizona.edu/software/uasoft.html.)

Because we are using data in a way that is not traditional for a differential equations course, in this article we will illustrate what we do more by example than exposition. These examples will involve questions given to students and comments on some possible solutions. We start with an example that could be used to introduce Newton's law of heating.

Example 1

The two columns in Table 1 give the temperature (T, measured in degrees Centigrade) of a person's finger as a function of time (t, measured in seconds). We want to develop a mathematical model which describes this data and allows us to find the temperature at other times.

(a) Plot the approximate numerical value of dT/dt (using central differences) as a function of time. This data appears approximately linear. If this is accurate, it would mean that the rise of temperature in this table is governed by an equation of the form

$$dT/dt = at + b,$$

where a and b are constants. Determine values of a and b which give a good fit to the data. What is the sign of a? Find the explicit solution of this differential equation and describe what happens to T as t approaches infinity. Is this a reasonable model for large values of time? (See Figures 1(a) and (b).)

Table 1

Time (seconds)	Temperature (degrees Centigrade)
0	20.0
2	22.8
4	24.2
6	25.6
8	26.8
10	27.5
12	28.2
14	28.8
16	29.5
18	29.9
20	30.4

(b) Plot the approximate numerical value of dT/dt as a function of temperature. This data appears approximately linear. If this is accurate, it would mean that the rise of temperature in this table is governed by an equation of the form

$$dT/dt = \alpha T + \beta,$$

where α and β are constants. Determine values of α and β which give a good fit to the data. What is the sign of α? Find the explicit solution of this differential equation and describe what happens to T as t approaches infinity. Is this a reasonable model for large values of time? (See Figure 2.)

(c) Expand the explicit solution obtained in part (b) in a Taylor series about $t = 0$. Explain the relationship between this Taylor series and the explicit solution obtained in part (a).

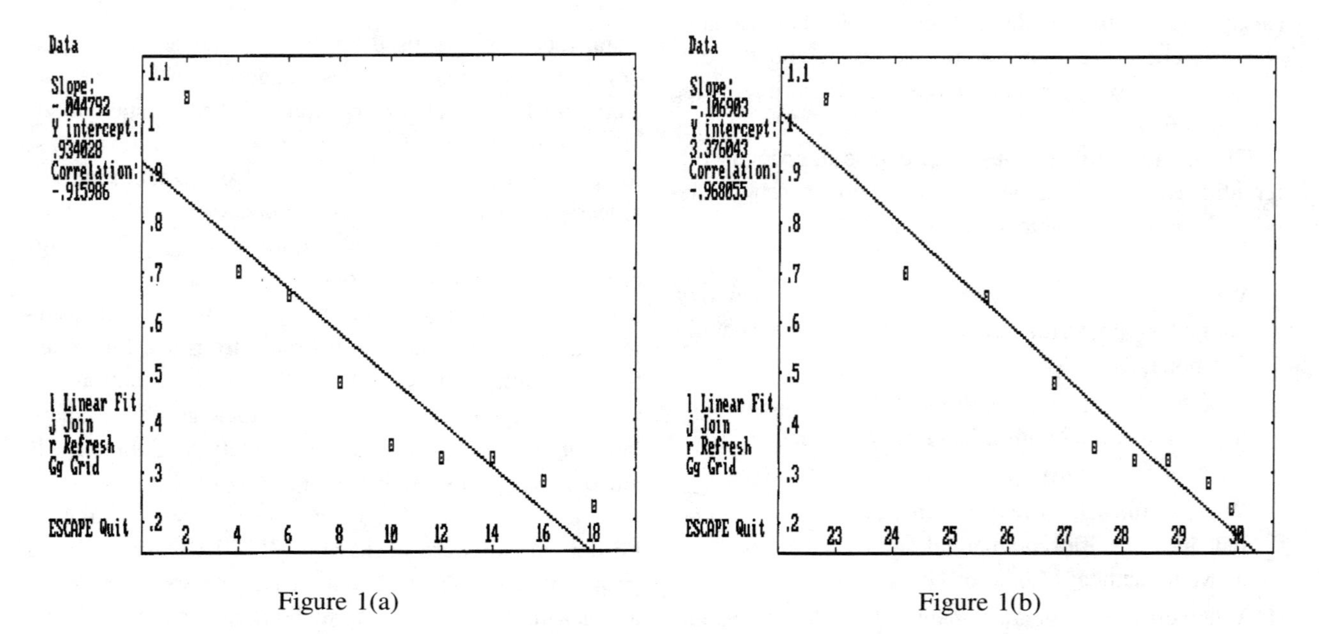

Figure 1(a)

Figure 1(b)

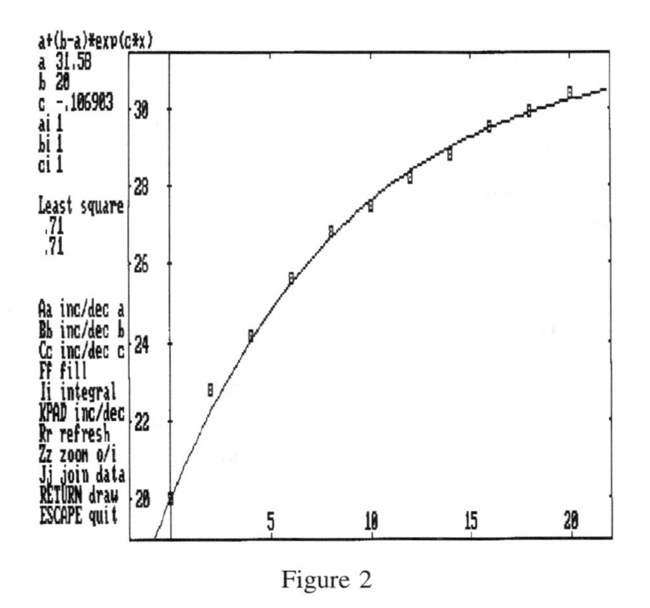

Figure 2

Comments on This Example The goodness of the linear fit in parts (a) and (b) is not very different (see Figure 1). However, the fact that a is negative in part (a) means that the model predicts a negative temperature for large values of time. A comparison of the results of part (b) is given in Figure 2. These data were collected from the finger of the author, so one conclusion drawn by his students was that it was a good model of a cold blooded instructor. Part (c) of this example gives a use of Taylor series in comparing two functions.

The next example concerns data where there may be some question as to what is the appropriate differential equation which describes the phenomena. It is built around data a local high school student collected as part of a special project. Because of the amount of time necessary to complete all parts of the stated problem, it is more typically used for group homework.

Example 2

A student shot a 120 grain hollow point bullet from a 300 Winchester Magnum rifle. He had equipment that measured the velocity, v, as a function of distance, x, and compiled the data given in Table 2. We are going to construct several mathematical models and analyze this data set in various ways.

(a) It is claimed that this data set looks linear, and if so would obey the law

$$v(x) = ax + b,$$

where a and b are constants. Find values of a and b so this linear model gives a good fit to the data set. With these values, estimate how far the bullet travels before coming to rest. Remembering that $v = dx/dt$, estimate the time when the bullet comes to rest.

Table 2. Velocity of bullet as a function of distance

Distance (ft)	Velocity (ft/sec)
0	3290
300	2951
600	2636
900	2342
1200	2068
1500	1813

Parts (b), (c), (d), and (e) repeat these questions for equations of the form $v(x) = ax^2 + bx + c$, $dv/dt = -kv^n$, $dv/dx = av + b$, and $dv/dx = ax + b$.

Comments on This Example In five of these models you can obtain values of the parameters which give a reasonable fit over the range of the given data. Thus other considerations are necessary in order to decide which model is most appropriate for predictions of what happens for distances greater than 1500 feet. In part (a) the velocity is zero, that is the bullet stops moving, when $x = -b/a$. However, the model predicts it will never reach this value in finite time! Two of the other models predict that as time increases, the bullet increases either in velocity or height. Most students really enjoy this problem.

The following example shows that solutions of the logistic differential equation is at variance with the data, while a simple modification provides an excellent fit.

Example 3

Nicotine Sulfate is introduced into a closed room containing mosquitos to determine how efficiently the chemical kills these insects. Table 3 shows the results of an experiment [1] where the proportion of mosquitos killed is given for various doses of Nicotine Sulfate. It is proposed that a possible model for this experiment is the logistic equation,

$$dy/dx = \alpha y(1 - y), \qquad (1)$$

where α is a specific constant, y is the proportion killed and x is the strength of the dose.

a) What is an appropriate initial condition for this problem? Use the explicit solution of this initial value problem to find the value of the parameter α which best fits this

Table 3. The efficiency of Nicotine Sulfate at killing mosquitos

Dose	Number in Room	Deaths	Efficiency
0.10	47	8	0.170
0.15	53	14	0.264
0.20	55	24	0.436
0.30	52	32	0.615
0.50	46	38	0.826
0.70	54	50	0.926
0.95	52	50	0.962

data. Are you satisfied with the fit you found? (Use this value in your explicit solution and compare the result with the data.)

b) Why is it reasonable that dy/dx be proportional to $1-y$ instead of $b-y$, with b not equal to 1?

c) Someone proposes that

$$dy/dx = \alpha x^{(-0.8)} y(1-y), \tag{2}$$

where α is a constant would be a better model for this situation. Find the explicit solution in this situation.

d) Use the data to find suitable choices for the constants in your solution. With these choices, plot your explicit solution on the same graph as the data. Which of the two differential equations provides a better model, (1) or (2)? Explain your reasoning.

The result of parts (c) and (d) appears in Figure 3.

Many students are interested in the use of differential equations to analyze and predict population growth. The typical model for such an application is either an exponential model or a logistic model. Many books use the population of the United States to illustrate time intervals over which each is a good model or a not so good model. Our next example involves the population of Kenya from 1950 to 1990.

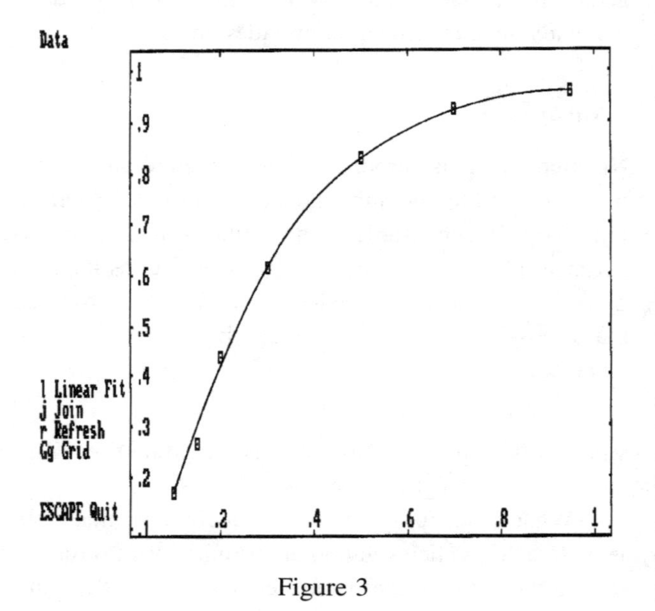

Figure 3

Example 4

Consider the population of Kenya in the years 1950–1990 as given in Table 4. If this population were growing exponentially it could be adequately modeled by the function $f(t) = 6.265e^{(\alpha t)}$, where is t represents the number of years after 1950 and α is suitably chosen.

Table 4. Population of Kenya

Year	Population (in millions)
1950	6.265
1955	7.189
1960	8.332
1965	9.749
1970	11.498
1975	13.741
1980	16.632
1985	20.353
1990	25.130

(a) Use technology to show that α cannot be determined to provide a very good fit to the data.

(b) Find the doubling times during this period. (i.e., estimate when the population is 12.5 million; 25 million). Are your results consistent with exponential growth?

(c) Show that a plot of $(1/P)dP/dt$ (using central differences) against t gives a straight line. Find values of the slope and intercept.

(d) Solve the differential equation from part (c) and plot the result on the same graph as the data in Table 4. Comment on the model's applicability for long periods of time.

Comments on Example 4 The solution from part (d) and the data are shown in Figure 4 and are a surprise to most students (and instructors). If this example is done in class, you could assign students a homework problem to find some city, state, or other country that was also growing at a rate greater than exponential, i.e., like $\exp(at^2)$.

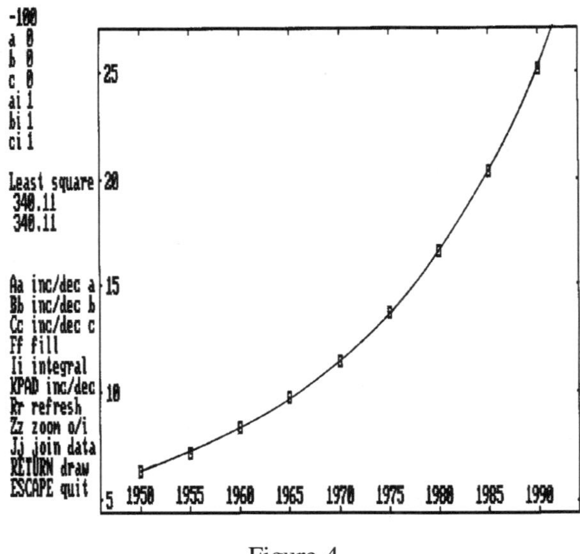

Figure 4

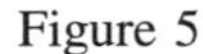
Figure 5

Example 5

Table 5 contains data from the United States Parachute Association that correspond to a sky diver in free fall in a stable spread-eagle position when falling from rest. It shows the sky diver's distance fallen (in feet) as a function of time (in seconds). The third column contains central difference quotients of the distance with respect to time.

(a) From Table 5 estimate the terminal velocity. Do you think this motion is described by a common model for falling bodies, $mdv/dt = mg$, where m and v are the mass and velocity of the diver, respectively, g is a gravitational constant and t is time? Explain your reasoning.

Table 5. Distance fallen (in feet) against time (in seconds)

Time (sec)	Distance (ft)	Difference Quotient
0	0	
1	16	31.0
2	62	61.0
3	138	90.0
4	242	114.0
5	366	131.0
6	504	143.0
7	652	152.0
8	808	159.5
9	971	165.0
10	1138	169.0
11	1309	172.5
12	1483	

(b) One modification to this differential equation is to include the effect of air resistance. This is often done by rewriting the equation in (a) as $mdv/dt = mg - kv^n$, where k and n are positive constants. (Because the velocity is always positive the way this problem was set up we do not need to worry about absolute value signs.) Determine the terminal velocity for this model. Do a phase line analysis to determine the stability of the equilibrium solution. Does your result agree with your common sense?

(c) What are the appropriate initial conditions regarding the diver's itinerary?

(d) Choose $n = 1$ and integrate to obtain the velocity and the distance fallen (recall $y = dv/dt$) as a function of time. (The parameter k/m will also appear in your solution.) Try various values of the terminal velocity in your solution and see if you can match the data in Table 5.

(e) Choose $n = 2$ and integrate to obtain the velocity and the distance fallen as a function of time. (Replacing m/k by V^2/g, where V is the terminal velocity simplifies some of the algebra.) Try various values of the terminal velocity in your solution and see if you can match the data in Table 1. Does this solution match the data better than the one in part (d)?

Comments on Example 5 Figure 5 shows the graph of the solution from part (e) along with the data. If students are interested, this example can lead to problems about the flight of ski jumpers, golf balls, baseballs, or badminton shuttlecocks [2, 3, 4, and 5].

The next example shows how more than one differential equation can be excellent models of some phenomena.

Table 6. Distance traveled by Carl Lewis as a function of time

Time (sec)	Distance (meters)
0.00	0
1.94	10
2.96	20
3.91	30
4.78	40
5.64	50
6.50	60
7.36	70
8.22	80
9.07	90
9.93	100

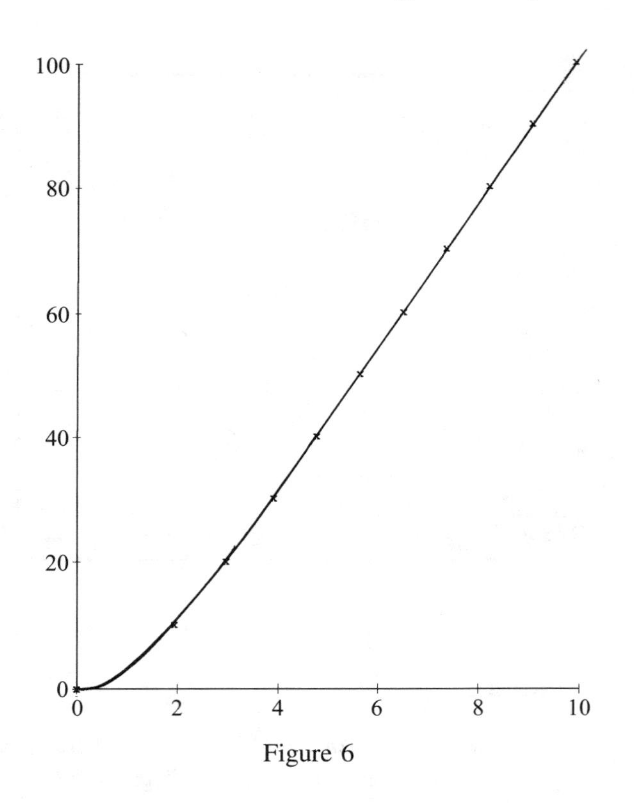

Figure 6

Example 6

Modeling a 100 meter race of Carl Lewis. The data in Table 6 is that of Carl Lewis in the 100 meter final of the World Championships in Rome in 1987 [6].

(a) Does it appear from the data that Carl Lewis attains a terminal velocity?

(b) If we model the distance that Carl Lewis runs as a function of time, what are the appropriate initial conditions?

(c) Does the differential equation $d^2y/dt^2 = a\exp(-bt)$ predict a terminal velocity? Integrate this equation subject to appropriate initial conditions to find this terminal velocity in terms of the parameters a and b.

(d) Integrate once more to find the distance as function of time. Obtain estimates of a and b from the data set and plot your resulting solution on the same graph as the data. Comment on the fit.

(e) Does the differential equation $d^2y/dt^2 = a - bdy/dt$ predict a terminal velocity? What is it (in terms of a and b)? Integrate this equation twice subject to appropriate initial conditions. Obtain estimates of a and b from the data set and plot your resulting solution on the same graph as the data. Comment on the fit.

The explicit solution from part(e) and the data from Table 6 are given in Figure 6.

Final Comments

Data from electric circuits or populations, such as from Ireland around the year of the potato famine, can be used in situations where you wish to piece together solutions of two differential equations.

Earlier in this paper some simple experiments are mentioned where students can collect their own data. All or parts of a class can be involved in experiments using commercial products for collection such as those distributed by IBM, Texas Instruments or Venier Software. Examples regarding Newton's law of cooling, a simple pendulum, a vibrating spring, and falling bodies are given in [7], along with other suggested projects.

Many other examples can be given which have application to other fields. For example, modeling the vapor pressure of toluene as a function of temperature may be of interest to a chemistry major while the allometric growth of fiddler crabs might appeal to a biology major. Other suggestions are the adoption of improved pasture management by ranchers to agriculture students, the learning response of rats to a psychology major, and why there are no very tall ballet dancers to fine arts majors. All of these examples, along with many, many others, may be found in [8] and [9]. Many of these may be downloaded from http://www.math.arizona.edu/~dsl/data/html which also contains links to other sources of data.

We have found that introducing data into our first course in ordinary differential equations has led students to be more interested in the applicability of the ODE's and therefore more willing to learn the material. It has also helped them to think about the reasonableness of the solution of an ODE. This will be of great value as the software provides more automatic differential equation solvers. There will always be cases for which they give misleading or incorrect answers.

References

1. Vazquez, J.O. (1995), Escuela Nacional de Ciencias Basicas, Mexico (personal communication).
2. True, E. (1993), "The Flight of a Ski Jumper" by E. True, C*ODE*E Newsletter, Spring pp. 5–8.
3. Erichson, R. (1983), "Maximum projectile range with drag and lift, with particular attention to golf," *American Journal of Physics*, **51**, pp. 357–362.
4. de Mestre, N. (1990), "The Mathematics of Projectiles in Sport," *Australian Mathematical Society Lecture Series* **6**, Cambridge University Press, Cambridge.
5. Peastrel, M., Lynch, R. and Armenti, Jr., A. (1980), "Terminal velocity of a shuttlecock in vertical fall," *The American Journal of Physics* **48**, pp. 511–513.
6. Pritchard, W. G. (1993), "Mathematical Models of Running," *SIAM Review*, **35**, pp. 359–379.
7. Lomen, D. O. , (1995), "Experiments with Probes in the Differential Equations Classroom," *College Math Journal*, **25**, pp. 453–457.
8. Lomen, D. and Lovelock, D. (1999), "Differential Equations: Graphics, Models, and Data," John Wiley & Sons, New York.
9. Lomen, D. and Lovelock, D. (1999), "Instructor's Resource Manual for Differential Equations: Graphics, Models, and Data," John Wiley & Sons, New York.

Qualitative Study of Differential Equations

V.S. Manoranjan
Washington State University

Abstract

This article is an introduction to dynamical systems in the context of differential equations, intended especially for mathematics, science and engineering majors who are interested in the direct applications of the subject. The treatment is kept at an intuitive and elementary level, but is designed to take the student from basic elementary concepts to the point where the exciting and fascinating advancements in the theory of nonlinear differential equations can be understood and appreciated. Topics such as phase plane, equilibrium points, periodic solutions, limit cycles and the associated stability ideas are treated informally and the students are introduced to investigative techniques useful in the application of differential equations to real-world problems.

Introduction

The study of ordinary differential equations is essential for students in many areas of science and technology. Many useful and interesting phenomena in engineering and life sciences which evolve continuously in time can be modeled by ordinary differential equations. So it is very important for the students to have a good understanding about ordinary differential equations, their solutions, and the different kinds of qualitative behavior the systems of ordinary differential equations can exhibit. In this chapter, we provide a brief qualitative study of ordinary differential equations making use of a variety of examples which show various types of qualitative behavior associated with ordinary differential equations. At this juncture, it should be pointed out that there are many excellent software packages that can generate the figures you will find in this chapter. In our opinion, it is a good idea to generate figures in the class itself, so that students can connect the theory with the visual objects and get a better conceptual learning experience. Since most of the real-world differential equation models are nonlinear, our focus will be on systems of nonlinear ordinary differential equations. Also, since this chapter is written at an introductory level, we restrict our presentation to second order systems.

Definitions

We begin by considering the system of ordinary differential equations (ODEs) given by

$$\frac{dx}{dt} = f(x,y) \quad \text{and} \quad \frac{dy}{dt} = g(x,y). \tag{1}$$

Here, we use t (for time, usually) to denote the independent variable; whereas, x and y are the dependent variables. The functions $f(x,y)$ and $g(x,y)$ are, in general, nonlinear and are continuously differentiable with respect to x and y.

Any functions $x(t)$ and $y(t)$ that satisfy (1) are a set of solutions, and a particular set of solutions can be found, provided one has specified values for x and y at a given time t. It is the usual practice to take the given time t to be zero, denote the specified values for x and y by $x(0)$ and $y(0)$, respectively, and call them the *initial values*.

If $f(x,y) = 0$ and $g(x,y) = 0$ simultaneously (i.e., the rates of change of x and y are both zero), for some $x = a$ and $y = b$, where a and b are constants, then the set $x(t) = a$ and $y(t) = b$ is the equilibrium solution to (1). In a geometrical representation, the point $(x(t), y(t)) \equiv (a,b)$ in the x-y plane is known as the *equilibrium point*.

The way we have written the system (1), the functions f and g do not involve t explicitly. Such a system is called an *autonomous system*. If we consider a system such as

$$\frac{dx}{dt} = f(x,y,t) \quad \text{and} \quad \frac{dy}{dt} = g(x,y,t) \tag{2}$$

where t appears in f and g explicitly, this system is called a *nonautonomous system*. However, by introducing a new dependent variable $z(= t)$, we can easily re-write (2) as

$$\frac{dx}{dt} = f(x,y,z)$$

$$\frac{dy}{dt} = g(x, y, z) \quad (3)$$
$$\frac{dz}{dt} = 1.$$

Now the system in (3) is third-order, but autonomous—i.e., we have arrived at an autonomous system from a nonautonomous system! However, one should note that this new autonomous system can never have an equilibrium solution, since the rates of change of x, y and z can not be zero simultaneously ($\frac{dz}{dt}$ is always 1). In this chapter, we confine our attention only to autonomous systems.

A particular solution to the autonomous system given in (1) can be thought of as a curve or trajectory in the two-dimensional Euclidean space $\mathbb{R}^2$ (i.e., the x-y plane). The points on this trajectory will have coordinates $(x(t), y(t))$ at any time t. The entire x-y plane will be full of interwoven trajectories, where every trajectory is a solution to some initial value problem. In fact, every point on a given trajectory can serve as initial value point for that trajectory.

Generally an arrow is placed on every trajectory to indicate the direction in which a point $(x(t), y(t))$ on the trajectory moves as t increases. Attached to every point $(x(t), y(t))$ is a vector with components $\{f[x(t), y(t)]\}$ and $\{g[x(t), y(t)]\}$ (corresponding to time rates of change of $x(t)$ and $y(t)$) and this vector is tangent to the trajectory at the point $(x(t), y(t))$ for every value of t. The vector $\{f(x, y), g(x, y)\}$ is said to specify *flow* in the $(x$-$y)$ plane and the plane is called the *phase plane* for the set of differential equations in (1). The task of finding solutions to the ODE system amounts to constructing curves (or trajectories) in the phase plane that are everywhere tangent to the flow vectors. The quest to understand the qualitative behavior of the system in (1) boils down, in concept, to the production of a composite portrait of all the trajectories of the flow (f, g). This is called a *phase portrait* and tells us everything interesting there is to know about all possible solutions of the ODE system.

Phase Plane Analysis

Phase plane analysis is a powerful process by which essentially complete qualitative information about all solutions of the ODE system such as (1) can be deduced by graphical methods coupled with the use of an appropriate linear system.

We will describe the various ideas associated with phase plane analysis using a variety of examples. First, let us consider a linear example

$$\begin{aligned} \dot{x} &= y \\ \dot{y} &= -x. \end{aligned} \quad (4)$$

The equilibrium points for this system can be determined by setting

$$\dot{x} = 0 \quad \text{and} \quad \dot{y} = 0.$$

Here, we obtain $x = 0, y = 0$ as the only equilibrium point. One could see (4) is none other than the simple harmonic oscillator equation $\ddot{x} + x = 0$, written as a first order system of ODEs. So, the solution of this simple harmonic oscillator is easily found as

$$x(t) = A\sin(t + \alpha) \quad \text{and} \quad y(t) = A\cos(t + \alpha) \quad (5)$$

where, A and α are constants. This means that $x(t)$ is a periodic solution which oscillates about the equilibrium state $x = 0$ with an amplitude A. Now, let us construct the phase portrait for this problem. The phase trajectories are the solutions of

$$\frac{dy}{dx} = \frac{\dot{y}}{\dot{x}} = \frac{-x}{y}, \quad (6)$$

which is separable, leading to

$$x^2 + y^2 = C, \quad \text{a constant,} \quad (7)$$

an equation of a circle with the center at the origin. (Note that from (5), $x^2 + y^2 = A^2\sin^2(t+\alpha) + A^2\cos^2(t+\alpha) = A^2$, constant, which is consistent with (7).) Therefore, the phase portrait consists of a family of concentric circles with the center at the origin $(0, 0)$, the equilibrium point (Figure 1). It should be noted that the direction of the trajectories can be determined by considering the signs of $\dot{x}$ and $\dot{y}$ in the four quadrants of the phase plane. An equilibrium point surrounded by closed paths (for example, circles in this case) in its immediate neighborhood is called a *center*. A center is a stable equilibrium point. In particular, a center is *neutrally stable*, meaning once displaced (even by a small amount) from it the system will never reach the equilibrium point again, but will oscillate about it with respect to time. Thus, when one encounters a phase portrait with a center, it means that the real system has an oscillating solution about that equilibrium point.

Now, let us consider a simple nonlinear example given by

$$\begin{aligned} \dot{x} &= xy \\ \dot{y} &= -x^2. \end{aligned} \quad (8)$$

For this example, we can easily see that every point on the y-axis is an equilibrium point. As in the earlier example,

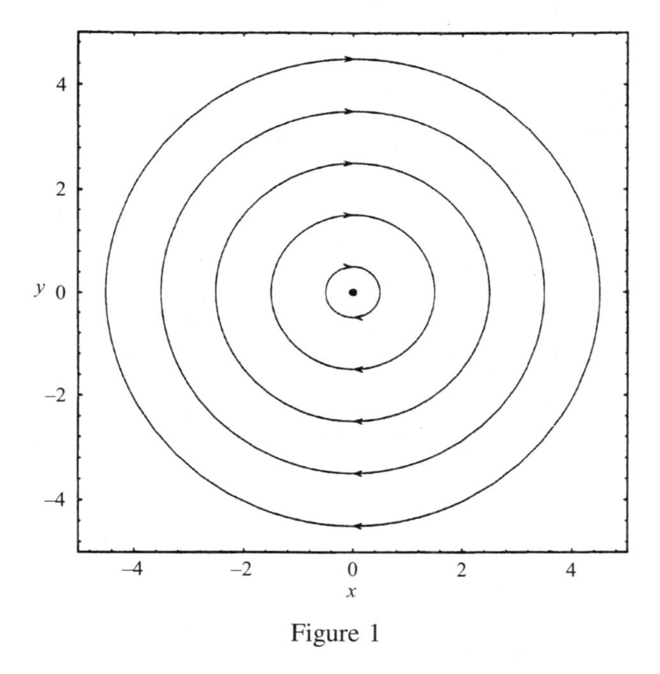

Figure 1

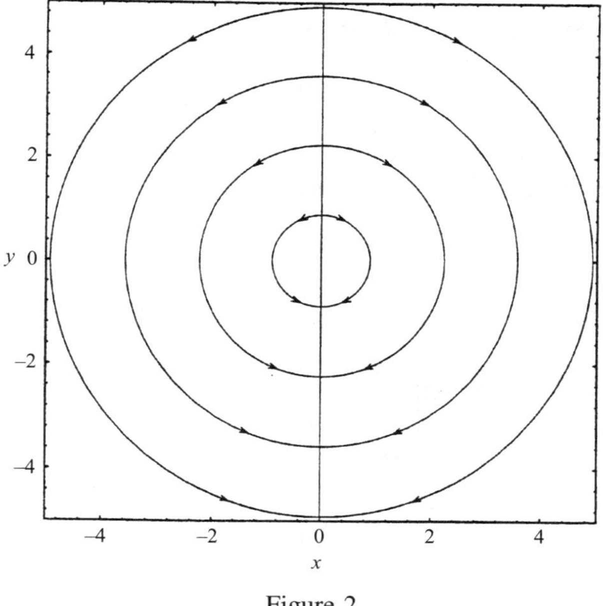

Figure 2

if we try to re-write (8) in terms of the unknown $x(t)$, we obtain the equation

$$x\ddot{x} - \dot{x}^2 + x^4 = 0. \tag{9}$$

Equation (9) is very nonlinear and it is not easy to solve this equation exactly. On the other hand, let us see whether we can construct the phase portrait for (8). As before, we can find the phase trajectories by solving

$$\frac{dy}{dx} = \frac{\dot{y}}{\dot{x}} = \frac{-x^2}{xy} = \frac{-x}{y}. \tag{10}$$

Of course, this again gives the concentric circles $x^2 + y^2 = C$, a constant. However, now, the direction arrows of the trajectories are different from that of the earlier example (see Figure 2). Since the equilibrium point $(0, 0)$ is surrounded by concentric circles, it is a center and has neutral stability. But what about the stability of every other equilibrium point on the y-axis? Let us look at the phase portrait in Figure 2. We see that the direction arrows are moving away—in either direction—from any point on the positive y-axis, meaning that any small displacement will move the system away from those equilibrium points. On the other hand, the direction arrows move towards any point on the negative y-axis, implying that whenever there is any displacement about an equilibrium point on that part of the y-axis, the system will come back to that equilibrium point. This means that the equilibrium points on the positive y-axis are unstable; but every equilibrium point on the negative y-axis is stable—moreover, it is *asymptotic stability*, since the direction arrows converge to the point.

Although in this simple nonlinear example we are able to construct the phase portrait without much trouble for most nonlinear systems, understanding the qualitative behavior near an equilibrium point is not that straightforward. One needs to employ the idea of *linearization* in order to get a magnified view of details in the neighborhood of an equilibrium point.

Linearization About An Equilibrium Point

We would like to find out what happens near an equlibrium point (x_0, y_0) for a system such as (1). So, let $\hat{x}$ and $\hat{y}$ be infinitesimal displacements from the equilibrium point and substitute $x = x_0 + \hat{x}$ and $y = y_0 + \hat{y}$ into (1). Now by expanding $f(x_0+\hat{x}, y_0+\hat{y})$ and $g(x_0+\hat{x}, y_0+\hat{y})$ as Taylor series about (x_0, y_0) and neglecting quadratic and higher order terms in $\hat{x}$ and $\hat{y}$, we obtain

$$\begin{aligned}
\frac{d}{dt}(x_0 + \hat{x}) &= f(x_0 + \hat{x}, y_0 + \hat{y}) \\
&= f(x_0, y_0) + \frac{\partial f}{\partial x}(x_0, y_0)\hat{x} \\
&\quad + \frac{\partial f}{\partial y}(x_0, y_0)\hat{y} \\
\frac{d}{dt}(y_0 + \hat{y}) &= g(x_0 + \hat{x}, y_0 + \hat{y}) \\
&= g(x_0, y_0) + \frac{\partial g}{\partial x}(x_0, y_0)\hat{x} \\
&\quad + \frac{\partial g}{\partial y}(x_0, y_0)\hat{y}.
\end{aligned} \tag{11}$$

Since (x_0, y_0) is an equilibrium point,

$$\frac{dx_0}{dt} = 0 = \frac{dy_0}{dt}$$

and $f(x_0, y_0) = 0 = g(x_0, y_0)$.

Therefore, (11) can be reduced to

$$\begin{aligned} \frac{d\hat{x}}{dt} &= a_{11}\hat{x} + a_{12}\hat{y} \\ \frac{d\hat{y}}{dt} &= a_{21}\hat{x} + a_{22}\hat{y}, \end{aligned} \tag{12}$$

where $a_{11} = \frac{\partial f}{\partial x}(x_0, y_0)$, $a_{12} = \frac{\partial f}{\partial y}(x_0, y_0)$, $a_{21} = \frac{\partial g}{\partial x}(x_0, y_0)$, and $a_{22} = \frac{\partial g}{\partial y}(x_0, y_0)$.

Now (12) is a linear system of equations which could be written as,

$$\frac{d}{dt}\begin{bmatrix} \hat{x} \\ \hat{y} \end{bmatrix} = \begin{bmatrix} a_{11} & a_{12} \\ a_{21} & a_{22} \end{bmatrix}\begin{bmatrix} \hat{x} \\ \hat{y} \end{bmatrix}$$

and the matrix

$$\begin{bmatrix} a_{11} & a_{12} \\ a_{21} & a_{22} \end{bmatrix}$$

is known as the *Jacobian matrix* associated with the nonlinear system (1). According to the linear theory, the general solution to a system such as (12), is a linear combination of exponential functions in time, where the exponents are given by the eigenvalues of the Jacobian matrix. Let us find these eigenvalues.

$$\det\begin{vmatrix} a_{11} - \lambda & a_{12} \\ a_{21} & a_{22} - \lambda \end{vmatrix} = 0$$

or

$$\lambda^2 - (a_{11} + a_{22})\lambda + (a_{11}a_{22} - a_{12}a_{21}) = 0.$$

Let $b = (a_{11} + a_{22})$ and $c = (a_{11}a_{22} - a_{12}a_{21})$. Then, the eigenvalues are given by,

$$\lambda_1 = \frac{1}{2}\left[b + \{b^2 - 4c\}^{\frac{1}{2}}\right]$$

and

$$\lambda_2 = \frac{1}{2}\left[b - \{b^2 - 4c\}^{\frac{1}{2}}\right].$$

Now we will consider the different possibilities for λ_1 and λ_2 and say what kind of qualitative behavior (12) will exhibit in the neighborhood of the equilibrium point (x_0, y_0). It should be pointed out that we will not look into special cases such as the case when both eigenvalues are zero, because for such a case linearization does not disclose anything. The general solution to (12) can be written as

$$\begin{bmatrix} \hat{x}(t) \\ \hat{y}(t) \end{bmatrix} = \alpha_1 \boldsymbol{s}_1 e^{\lambda_1 t} + \alpha_2 \boldsymbol{s}_2 e^{\lambda_2 t} \tag{13}$$

where $\boldsymbol{s}_1$ and $\boldsymbol{s}_2$ are the eigenvectors corresponding to λ_1 and λ_2 respectively.

Case 1: λ_1 and λ_2 are both real and negative—i.e., $b < 0, c > 0$ and $(b^2 - 4c) > 0$.

In this case, it is obvious from (13) that the solution will decay exponentially to zero. This says that the infinitesimal displacements will die out and the nonlinear system will return to the equilibrium state (x_0, y_0). So, (x_0, y_0) is a stable equilibrium point and it is called a *stable node*.

Case 2: λ_1 and λ_2 are both real and positive—i.e., $b > 0, c > 0$ and $(b^2 - 4c) > 0$.

This case is the exact opposite of the earlier case. Here the solution (13) will grow exponentially and therefore the nonlinear system will not return to the equilibrium state (x_0, y_0). So we have an unstable equilibrium point (x_0, y_0), known as an *unstable node*.

Case 3: λ_1 and λ_2 are both real with one eigenvalue being positive and the other, negative—i.e., $c < 0$.

Here, a portion of the solution in (13) grows exponentially in time while the other portion decays exponentially in time. In this case (x_0, y_0) is known as a *saddle point* and it is an unstable equilibrium point.

Case 4: λ_1 and λ_2 are complex conjugates with a negative real part—i.e., $b < 0, c > 0$ and $(b^2 - 4c) < 0$.

Now, because of the imaginary parts in the eigenvalues, the solution (13) will be oscillatory. However, the negative real part will make the solution decay exponentially at the same time. So, what we get is an oscillating solution which decays exponentially. In the phase plane, this solution will correspond to a trajectory that spirals into the point (x_0, y_0). In this case, the equilibrium point (x_0, y_0) is stable and it is called a *stable spiral point*.

Case 5: λ_1 and λ_2 are complex conjugates with a positive real part—i.e., $b > 0, c > 0$ and $(b^2 - 4c) < 0$.

This is the opposite of the prior case and the solution (13) will be an oscillating solution which grows exponentially. So, the corresponding phase trajectory will spiral out of the equilibrium point (x_0, y_0). Hence the equilibrium point is unstable and is called an *unstable spiral point*.

Case 6: λ_1 and λ_2 are complex conjugates and purely imaginary—i.e., $b = 0, c > 0$, and $(b^2 - 4c) < 0$.

Now the solution (13) will be just oscillatory—a periodic function in time—without any components that either grow or decay. Therefore, as seen before in an example, the phase trajectories will be concentric closed curves around the equilibrium point. The equilibrium point (x_0, y_0) is known as a *center* and is *neutrally stable*.

In the following section, we will look at a few examples, which illustrate some of these cases. It should be emphasized that the linearization method gives the qualitative behavior only in the *locality* of the equilibrium points. Therefore at times one might encounter limitations to the method when applied to certain nonlinear systems.

Illustrative Examples

In this section, we present some nonlinear systems of ODEs with interesting and varied qualitative behavior. For every example, we will determine the equilibrium points and analyze their stability. The corresponding phase portraits are presented in Figures 3–7, and were produced using *Mathematica*.

(a)

$$\begin{aligned} \dot{x} &= -y - x\sqrt{x^2+y^2} \equiv f(x,y) \\ \dot{y} &= x - y\sqrt{x^2+y^2} \equiv g(x,y) \end{aligned} \qquad (A)$$

By setting $\dot{x} = 0 = \dot{y}$, we find, the only equilibrium point to be (0,0). Let us analyze the stability of this equilibrium point using linearized analysis. In general, the

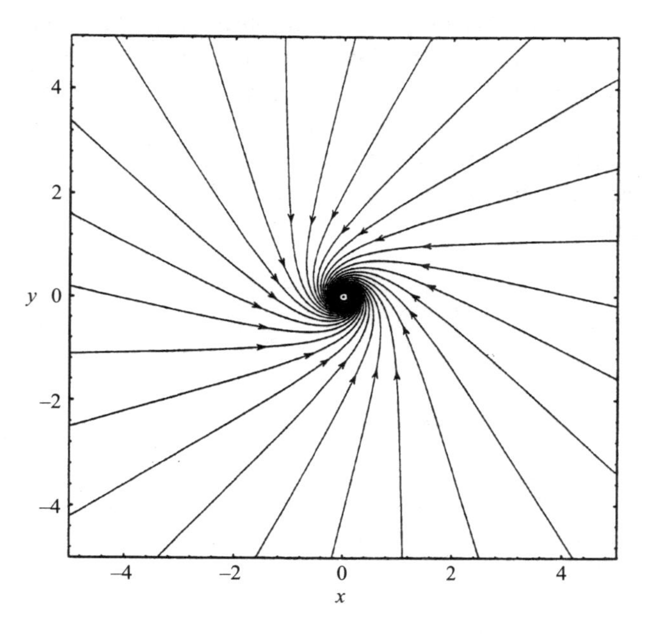

Figure 3

associated linearized system of ODEs is given by

$$\begin{bmatrix} \dot{x} \\ \dot{y} \end{bmatrix} = \begin{bmatrix} \dfrac{\partial f}{\partial x} & \dfrac{\partial f}{\partial y} \\ \dfrac{\partial g}{\partial x} & \dfrac{\partial g}{\partial y} \end{bmatrix}_{(0,0)} \begin{bmatrix} x \\ y \end{bmatrix}. \qquad (A1)$$

Here we have

$$\begin{bmatrix} \dot{x} \\ \dot{y} \end{bmatrix} = \begin{bmatrix} 0 & -1 \\ 1 & 0 \end{bmatrix} \begin{bmatrix} x \\ y \end{bmatrix}.$$

The eigenvalues of the Jacobian matrix $\begin{bmatrix} 0 & -1 \\ 1 & 0 \end{bmatrix}$ are found as

$$\lambda_{1,2} = \pm i, \quad \text{where } i = \sqrt{-1}.$$

Therefore, since the eigenvalues are purely imaginary, we will have neutral stability and $(0,0)$ is a center. So we would expect closed concentric phase trajectories surrounding $(0,0)$. However when the phase portrait was drawn using *Mathematica* we get Figure 3, which shows that $(0,0)$ is *not* a center, but a stable spiral point! So, what has gone wrong with our analysis?

Let us look at the equations given in (A) again. Using the polar coordinates (r, θ) and the relationships

$$r^2 = x^2 + y^2 \quad \text{and} \quad \tan\theta = y/x,$$

we can re-write (A) as

$$\begin{aligned} \dot{r} &= -r^2 \\ \dot{\theta} &= 1\ . \end{aligned} \qquad (A2)$$

From the first equation in (A1), since $\dot{r} < 0$, it is clear that r decreases with time. This shows that the phase trajectory will spiral into $(0,0)$, implying a stable spiral point at the origin. So, we are in agreement with the phase portrait presented in Figure 3.

This example demonstrates the limitation of the linear analysis. Even though the linear analysis showed the equilibrium point to be a center, the reality turned out to be different. One should always remember that the linear analysis gives only the local behavior.

(b)

$$\begin{aligned} \dot{x} &= y + 5x(x^2+y^2)^{1/2} \\ \dot{y} &= -x + 5y(x^2+y^2)^{1/2} \end{aligned} \qquad (B)$$

The equilibrium point for this system is $(0,0)$. By performing linear analysis, it can be shown that $(0,0)$ is a center which is neutrally stable. So, because of the previous example, is it safe for us to assume that $(0,0)$ is

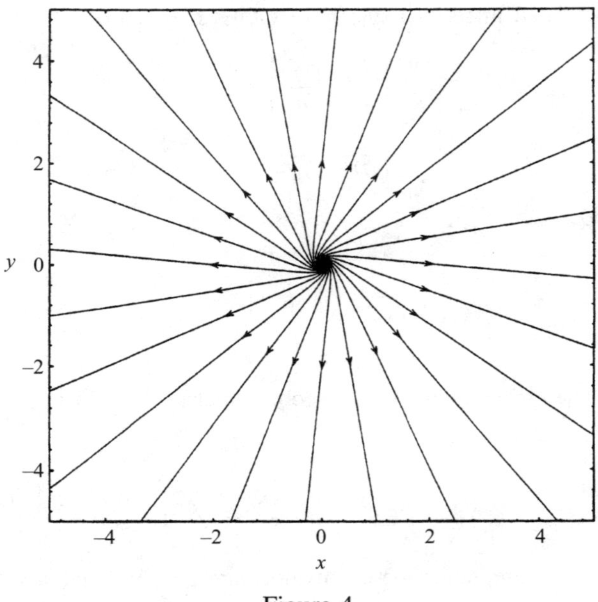

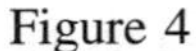
Figure 4

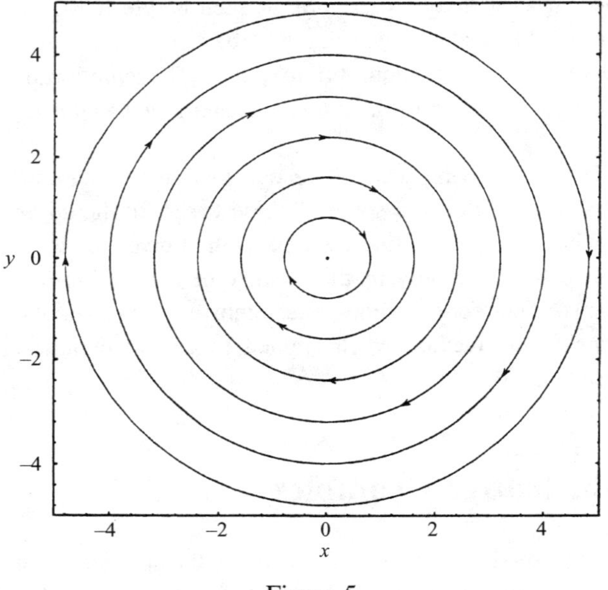

Figure 5

really a stable spiral point? Let us find out. If we re-write the equations in (B) in terms of polar co-ordinates (r, θ), we obtain

$$\begin{aligned} \dot{r} &= 5r^2 \\ \dot{\theta} &= -1 \ . \end{aligned} \qquad (B1)$$

Now, in (B1), $\dot{r}$ is always positive. This means that r increases with time and so, the phase trajectory will spiral out of $(0, 0)$. Therefore, $(0, 0)$ is an unstable spiral point (Figure 4). This clearly demonstrates that whenever the linear analysis determines an equilibrium point to be a center, one should not automatically accept that result to be valid for the full nonlinear system.

(c)

$$\begin{aligned} \dot{x} &= y + y\sqrt{x^2 + y^2} \\ \dot{y} &= -x - x\sqrt{x^2 + y^2} \end{aligned} \qquad (C)$$

Again for this system, the only equilibrium point is (0,0) and the linear analysis shows that this point is a center. From the two examples above, we know now that linear analysis has limitations and one has to be careful about determining the qualitative nature of the equilibrium point with respect to the nonlinear system. As before, let us transform the system in (C) into polar coordinates (r, θ). The resulting system will be

$$\begin{aligned} \dot{r} &= 0 \\ \dot{\theta} &= -1 - r \ . \end{aligned} \qquad (C1)$$

Since $\dot{r} = 0$ in (C1), that will lead to $r =$ const., which are concentric circles (i.e., closed phase trajectories) with center at the origin (Figure 5).

So, here $(0, 0)$ will still be a center even for the nonlinear system! This further demonstrates that one needs to be really careful whenever the linear analysis predicts an equilibrium point to be a center.

(d)

$$\begin{aligned} \dot{x} &= x - y \\ \dot{y} &= 1 - xy \end{aligned} \qquad (D)$$

For this example there are two equilibrium points $(-1, -1)$ and $(1, 1)$. If one performs linear analyses near both equilibrium points, it is easy to show that $(-1, -1)$ is an unstable spiral point (eigenvalues of the Jacobian matrix are complex conjugates with positive real part) and $(1, 1)$ is a saddle point (real eigenvalues of opposite signs). The phase portrait constructed by *Mathematica* (Figure 6) shows that the determinations made here using the linear analysis are true even for the nonlinear system.

(e)

$$\begin{aligned} \dot{x} &= y + x(x^2 + y^2)\sin\left\{\frac{\pi}{\sqrt{x^2 + y^2}}\right\} \\ \dot{y} &= -x + y(x^2 + y^2)\sin\left\{\frac{\pi}{\sqrt{x^2 + y^2}}\right\} \end{aligned} \qquad (E)$$

In order to analyze this system, we will re-write it in polar coordinates (r, θ). Then, we obtain

$$\begin{aligned} \dot{r} &= r^3 \sin\left\{\frac{\pi}{r}\right\} \\ \dot{\theta} &= -1 \end{aligned} \qquad (E1)$$

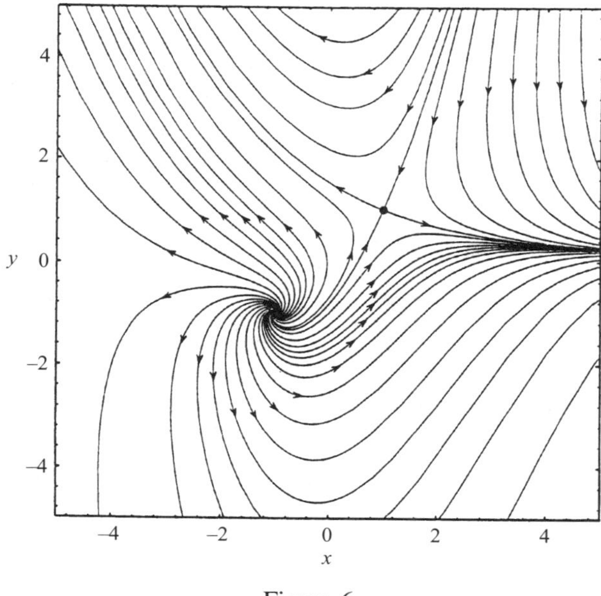

Figure 6

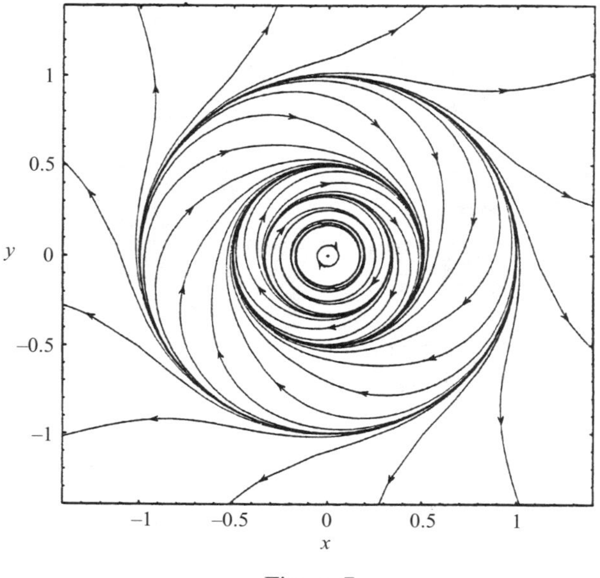

Figure 7

Now, $\dot{r} = 0$ will give either $r = 0$, the origin or

$$\frac{\pi}{r} = k\pi, \quad \text{where } k \text{ is an integer.}$$

In other words, either $r = 0$ or $r = \frac{1}{k}$, k, an integer.

So $(0, 0)$ is an equilibrium point and there are closed phase trajectories (circles of radii $1/k$), which are known as *limit cycles*. These limit cycles correspond to periodic solutions in time. But between any two limit cycles given by $r = \frac{1}{k+1}$ and $r = \frac{1}{k}$, $\dot{r} \neq 0$. For example, if we pick $r = \frac{1}{k+1/2}$, then we have,

$$\dot{r} = r^3 \sin\{(k + 1/2)\pi\}. \tag{E2}$$

If k is odd, then $\sin\{(k + 1/2)\pi\} = -1$ and so $\dot{r} < 0$ implying a decreasing spiral. Whereas, when k is even, $\sin\{(k + 1/2)\pi\} = 1$ and $\dot{r} > 0$, giving an increasing spiral. This also means that a limit cycle with radius $1/k$, where k is an even integer, is a stable limit cycle, while the one where k is an odd integer is an unstable limit cycle. The phase portrait for this example will consist of an equilibrium point at the origin, limit cycles of radii $\frac{1}{k}$ (k, an integer), and spirals in between limit cycles, which alternate between spiralling inwards or outwards based on whether k is odd or even. We present this portrait in Figure 7.

Conclusion

In this chapter, we have presented a brief introduction to the study of nonlinear differential equations. We have attempted to introduce the reader to some of the basic concepts in dynamical systems in an informal manner and through a variety of examples by making use of *Mathematica* to construct the associated figures. It would be a mistake to assume that this chapter presents a comprehensive study of known and practically useful results for nonlinear dynamical systems. For a more rigorous approach to nonlinear differential equations and to learn about various other useful and interesting concepts, the reader is advised to consult books such as, Chow & Hale (1982), Coddington & Levinson (1955), Grimshaw (1990), Guckenheimer & Holmes (1983) and Jordan & Smith (1989). In fact, in writing this chapter we have found these books very useful.

References

1. Chow, S.N., and Hale, J.K., Methods of bifurcation theory, Springer-Verlag, 1982.
2. Coddington, E.A., and Levinson, N., Theory of ordinary differential equations, McGraw-Hill, 1955.
3. Grimshaw, R., Nonlinear ordinary differential equations, Blackwell Scientific Publications, 1990.
4. Guckenheimer, J., and Holmes, P., Nonlinear oscillations, dynamical systems and bifurcations of vector fields, Springer-Verlag, 1983.
5. Jordan, D.W., and Smith, P., Nonlinear ordinary differential equations, Clarendon Press, 1989.

Teaching Numerical Methods in ODE Courses

Lawrence F. Shampine
Ian Gladwell
Southern Methodist University

Abstract

Nowadays, numerical methods play an important role in courses on ordinary differential equations (ODEs). We believe that popular texts place far too much emphasis on the derivation and study of numerical methods *per se*. Because the aim is to teach students about ODEs, the emphasis should be placed on teaching what they need to know so as to use quality software to gain insight about solutions and their behavior. Here we present briefly the fundamentals of numerical methods from this perspective. The approach implies that certain topics in the theory of ODEs ought to be discussed even in a first course.

We describe briefly quality software for initial value problems available in several distinct areas of application. In this we comment about how the area can influence the techniques used. Pointers are also provided to software for boundary value problems.

1. Introduction

Numerical methods for ordinary differential equations (ODEs) is perhaps the most difficult of the "elementary" topics of numerical analysis. Almost all undergraduates study ODEs before numerical methods, if they study numerical methods at all. And, it is hard to develop a good understanding of numerical techniques for ODEs without the support of other, more fundamental, topics in numerical computation. Perhaps for this reason, it is common that ODE texts devote a chapter to the development of some simple formulas for initial value problems (IVPs) and illustrate their use with some computations with constant step size. We believe the emphasis given to this material is misplaced: it is a digression from the point of the course, which is to gain understanding of the solutions of ODEs and their behavior; it is not helpful to those who do not later solve ODEs numerically; and it is a superficial introduction to the topic that may mislead those who later do.

What are our credentials? Both of us have been involved in the development of ODE software and the accompanying theory for 25 years, i.e., nearly from the inception of general-purpose software for ODEs. With our associates at a number of institutions, we have written some of the most widely used ODE software. Our programs are found in the major software libraries of IMSL [25], NAG [30], SLATEC [15] and the collection Netlib [13]. They are foundation codes in application software such as the computing environments MATLAB [40] and Maple [11], the CSEP project [1], and teaching packages such as that of Polking [33] and ODE Architect [31]. We are still actively developing software with, e.g., the first author leading the development of the MATLAB ODE Suite [38] and the second involved in the ODE software used by the TI-85 [29] and TI-92 calculators. In addition to developing software, recent publications include a monograph [35] on the practical numerical solution of IVPs and a paper [36] on the historical development of IVP software based on explicit Runge-Kutta methods. In addition to our experience in research and development into numerical methods, we have taught courses on ODEs and on numerical methods at all levels.

We begin with a brief presentation of the fundamentals of the numerical solution of ODEs from the perspective of a user (the teacher) who must teach other users (the students) how to employ the software and how to interpret the results. It is not intended that this material be used directly in the classroom, rather that it inform the teacher who should thus be better positioned to explain what to expect of numerical software without having to teach numerical methods *per se*. To appreciate what numerical ODE software can contribute to a course on ODEs, it is necessary to understand certain topics in the theory of differential equations, which we identify. We have had a long-term interest in boundary value problems (BVPs), see e.g. [3, 4, 18, 20]. Because of our experience dealing with BVPs arising in practice and because of the light a study of such problems can shed on IVPs, we believe that

ODE courses would benefit from more attention devoted to BVPs.

We conclude with pointers to some quality software, along with comments about how developments in general scientific computation are percolating into contexts closer to the classroom, specifically computing environments, teaching packages, and calculators. We do not aim to be exhaustive, so omission of a mention of a piece of software should not necessarily be viewed as negative.

2. Can it be solved numerically?

The reader of popular ODE texts may well be misled in two ways about the usefulness of numerical methods for ODEs. One is in thinking that *all* ODEs can be solved numerically, and the other in thinking that analysis and insight are not important in the practical solution of ODEs.

General-purpose numerical software (and the formulas which underlie it) is not suitable for certain kinds of problems typically studied in a first course. Numerical software invariably assumes that the functions defining the ODE are rather smooth. It is true that robust software may cope with isolated points where the solution is not smooth, but it is imprudent to rely upon this. Often modellers make use of piecewise constant or linear functions so as to obtain equations that can be solved analytically. For example, a sawtooth forcing function is commonly seen in some applications; another example appears in Section 7. Such forcing functions may allow us to solve the model analytically, but they can be disastrous if we try to solve the model numerically.

Problems with singular coefficients preclude standard software. To solve such problems it is necessary first to deal analytically with the solution in the neighborhood of a singular point and then to apply the code at some (small) distance from the singularity. So you must analyze the problem even when you intend from the beginning to solve it numerically. However, from our practical experience, we believe that the material covered in a first ODE course and its emphasis should be rather different from what appears to be standard. With the exception of evaluating special functions, it is rare to use more than a few terms in a series solution of a real problem. Much more emphasis should be given to computing the leading term (or even just its form) to understand the possible behavior of solutions near the singularity. Two terms are usually sufficient for numerical purposes. The first provides an approximate solution a distance h away from the singularity and the relative size of the second provides some idea of how small h must be to obtain an acceptable approximation. Computing a couple of terms ought to be illustrated for problems much more general than the scalar, second order linear equations with simple coefficients taken up in the typical introductory text. In particular, some attention ought to be given to nonlinear problems and to systems of equations.

Standard software accepts only problems posed on finite intervals, hence excludes many problems taken up in ODE courses where attention is often directed at long-term behavior. For that matter, standard methods cannot be relied upon for accurate solutions on "long" time intervals. Laplace and other transform methods are quite effective for a class of problems that simply cannot be resolved satisfactorily by standard software because the interval is long and/or the functions are not smooth. Also, Laplace transforms deal gracefully with some problems with periodic coefficients. Software for BVPs in general and periodic problems in particular is much less readily available than for IVPs, so is much less well-known.

We must expect numerical methods to make errors, and the most we can hope of a code is that the errors are "small". Obviously it is not possible to solve numerically problems with solutions that are very sensitive to changes in the data of the problem. The most extreme case is a problem with more than one solution. It is clear that we cannot hope to solve such a problem without special preparation to identify and at least start the integration on the desired solution. The only large class of problems with non-unique solutions for which an analysis of the problem has permitted the development of relatively general software is the Sturm-Liouville eigenvalue problem.

Problems with solutions that "blow up", like $y' = 1 + y^2$, $y(0) = 0$ for which $y(x) = \tan(x)$, or approach the boundary of the region where the equation is defined, like $y' = \sqrt{1 - y^2}$, $y(0) = 0$ for which $y(x) = \sin(x)$, are common. It is reasonable to ask that a code track such a solution to the boundary, but difficulties near the boundary are to be expected. In the first example, eventually the numerical solution will be too large for the computer arithmetic. A code might quit even earlier because the step size becomes too small to advance the integration in finite precision arithmetic. In the second example, we must expect trouble near $\pi/2$ because even tiny numerical errors might lead to an approximation to the solution for which it is impossible to evaluate the square root. In either case, it will be necessary to resort to analytical means to study the solution near the boundary, if it is of interest there. These are typical of the difficulties arising in the solution of practical BVPs, of which more later. We should inform students that for a sufficiently smooth differential equation, an IVP will have a unique solution that extends to a boundary where smoothness is lost and that numerical methods can be applied in the same domain with some hope of success, at least until this boundary is approached.

3. Getting ready to solve it.

Typical introductory ODE courses concentrate on a single equation of first or second order, often linear. This is understandable for learning solution techniques and what kind of behavior is possible, but it is not at all representative of applications. Problems typically arise as systems of nonlinear equations of mixed orders. The systems might be large–in some applications *thousands* of equations are common. General-purpose numerical software expects problems to be presented in a standard form, namely

$$\begin{aligned} y' &= f(x,y),\ a \le x \le b, \\ y(a) &= A;\ x \in R,\ y, f \in R^n. \end{aligned} \tag{1}$$

It is of the utmost practical importance that students learn how to formulate problems in this way. After all, if they cannot set up the problem in a form acceptable to the codes, they cannot solve it.

There is a good reason why there are very few codes that accept problems of the general form $F(x, y, y') = 0$, the same reason that such problems are not much discussed in courses on ODEs; existence and uniqueness of solutions is problematical. The equation $(y')^2 + 1 = 0$ makes the point about existence. The situation is entirely different for the standard form (1) because for such problems just a little smoothness of $f(x, y)$ implies existence and uniqueness. Formulating a more general problem in this standard form for its numerical solution forces us to confront the questions of existence and uniqueness. For instance, the equation $(y')^2 - 1 = 0$ requires us to decide whether we want a solution that satisfies $y' = +1$ or one that satisfies $y' = -1$ or a combination of the two.

It would suffice just to show students how to introduce variables so as to formulate a problem as a system of first order equations, but an example makes the point that there is more than one way to prepare a problem for its numerical solution. This example is taken from [35] where other aspects of the task are considered. Suppose we wish to solve $(p(x)y')' + q(x)y = g(x)$, $p(x) > 0$, an equation that arises in many physical contexts. One way to proceed is first to write the equation as

$$y'' + \frac{p'(x)}{p(x)}y' + \frac{q(x)}{p(x)}y = \frac{g(x)}{p(x)}$$

and then to introduce variables $y_1 = y$, $y_2 = y'$ that satisfy the first order system

$$\begin{aligned} y_1' &= y_2 \\ y_2' &= -\frac{p'(x)}{p(x)}y_2 - \frac{q(x)}{p(x)}y_1 + \frac{g(x)}{p(x)} \end{aligned}$$

This formulation requires the derivative $p'(x)$, which may be inconvenient or impossible to obtain. Moreover, many physical problems involve functions $p(x)$ that are only piecewise smooth. For such problems the flux $p(x)y'(x)$ must be continuous at points where $p(x)$ has a jump, implying that $y_2(x)$ must have a corresponding jump there. A better formulation is to introduce variables $z_1 = y$, $z_2 = p(x)y'$ that satisfy the first order system

$$\begin{aligned} z_1' &= \frac{z_2}{p(x)} \\ z_2' &= -q(x)z_1 + g(x) \end{aligned}$$

With this form we do not need to differentiate the coefficient $p(x)$ and the solution variables are continuous everywhere, simplifying numerical integration of the equation. There are interesting problems of this form for which $p(x) \to 0$ at an end point, particularly the initial point. The equation is singular at such a point, requiring some analysis before integration. This analysis is less involved in the second formulation.

In addition to (1), another standard form of some importance for IVP software is the "special" second order system

$$\begin{aligned} y'' &= f(x,y),\ a \le x \le b, \\ y(a) &= A,\ y'(a) = A';\ x \in R,\ y, f \in R^n. \end{aligned} \tag{2}$$

Such problems are given special attention because they arise widely in practice, e.g., in the equations of motion for the multibody systems of astronomical and molecular dynamics, and because there are special methods for (2) that are considerably more efficient than a standard method applied to an equivalent first order system of the form (1).

It makes no difference to quality software whether equations are complicated and, in particular, whether they are linear. With modern computing systems and quality software, the cost of solution does not depend strongly on the number of equations. Unfortunately students often do not appreciate these facts because of the emphasis given to very simple equations that permit analytical investigation. We ought to introduce them to more realistic problems even in a first course. An example we have used in the classroom exploits the fact that the teaching package MDEP [10] allows scalar equations of orders up to four. When we introduce the students to the software, we ask them to integrate the Blasius equation $\eta''' = -\eta\,\eta''/2$ with initial conditions $\eta(0) = 0$, $\eta'(0) = 0$. The problem actually arises as a BVP for which an $\eta''(0)$ is to be found such that $\eta'(x) \to 1$ as $x \to \infty$. It is possible to use a group of transformations to convert the BVP to an IVP, but it suffices for present purposes simply to say that it has been determined computationally that $\eta''(0) \approx 0.332$. The students are asked to confirm that this initial condition results in the desired behavior for η "large". They also confirm that small changes in this initial value lead to small changes in the numerical solution. This problem has a number of virtues:

- The physical origin of the problem is easily understood.
- The ODEs are very simple so that coding them up for numerical solution is easy and, as it happens, the integration is inexpensive.
- The solution is important as a building block in modelling more complicated fluid flows where scientists must compute it regularly by numerical means.
- The equation is not amenable to analysis at this level but, as indicated above, more advanced analysis can yield useful results.

4. How do you solve it?

The typical IVP code accepts problems of the form (1). The underlying numerical method is a discrete variable method, meaning that the method produces approximations $y_j \approx y(x_j)$ on a mesh $a = x_0 < x_1 < \cdots < x_N = b$. The computation starts with the known vector $y_0 = y(x_0) = A$ and, for $j = 0, 1, \ldots, N-1$, advances from (x_j, y_j) one step of length $h_j = x_{j+1} - x_j$ to obtain an approximation y_{j+1} at x_{j+1}. Although some popular codes produce results only at mesh points, many recent codes approximate the solution for all x, filling in values between mesh points by means of a "continuous extension". Details about the formula(s) used for taking a step and the continuous extension are not important to a course on ODEs, but some understanding of these matters facilitates an appropriate use of the codes.

Taylor series methods are easily understood and illustrate many characteristics typical of popular methods. The first point to be appreciated is that on reaching x_j, a code has available y_j, not $y(x_j)$. For this reason it tries to approximate the solution $u(x)$ of the ODE with initial condition $u(x_j) = y_j$, the "local solution". Later we discuss the implications of this key fact. The idea is to expand the local solution in a Taylor series

$$\begin{aligned} u(x_j + h) &= u(x_j) + h_j u'(x_j) + \frac{{h_j}^2}{2!} u''(x_j) + \cdots \\ &\quad + \frac{{h_j}^k}{k!} u^{(k)}(x_j) + \cdots \end{aligned}$$

In principle the derivatives here can be obtained from $f(x, y)$:

$$\begin{aligned} u'(x_j) &= f(x_j, y_j) \\ u''(x_j) &= \frac{\partial f}{\partial x}(x_j, y_j) + \frac{\partial f}{\partial y}(x_j, y_j) u'(x_j) \\ &= f_x(x_j, y_j) + f_y(x_j, y_j) f(x_j, y_j) \\ &\ldots \end{aligned}$$

The first derivative is readily available, but the presence of the Jacobian matrix f_y makes evaluating even the second derivative complicated. A Taylor series method defines an approximation y_{j+1} by means of a fixed number of terms from the series. The simplest Taylor series method is

$$y_{j+1} = y_j + h_j f(x_j, y_j),$$

commonly known as Euler's method. A more accurate method, obtained by keeping another term in the Taylor series, is

$$\begin{aligned} y_{j+1} &= y_j + h_j f(x_j, y_j) \\ &\quad + \frac{h_j^2}{2!}[f_x(x_j, y_j) + f_y(x_j, y_j) f(x_j, y_j)]. \end{aligned} \tag{3}$$

The accuracy of a Taylor series method depends on the number of terms and the size of h_j. The method of order k drops all terms in the tail of the series beginning with ${h_j}^{k+1} u^{(k+1)}(x_j)/(k+1)!$. If h_j is sufficiently small, terms decrease rapidly in the expansion and the error $u(x_{j+1}) - y_{j+1}$ is approximately equal to the first term dropped, which is to say $O({h_j}^{k+1})$. All this presumes that the solution is sufficiently smooth. It is typical that the higher the order, the more smoothness is needed. If $H = ma_j x h_j$, a method of order k makes an error at each step that is uniformly $O(H^{k+1})$. If the steps are not allowed to vary too much in the course of the integration, a code will reach b in $O(H^{-1})$ steps. Indeed, with a constant step size, it would take $(b-a)/H$ steps. Roughly speaking, for a stable method the errors do no worse than add up so that the true error $y(x_j) - y_j$ is uniformly $O(H^k)$, justifying the name "order k". In particular, Euler's method and the method (3) are convergent of order 1 and 2 respectively.

All production codes vary the step size automatically so as to approximate $u(x_{j+1})$ to a specified accuracy. Obviously this requires that the codes be able to estimate the accuracy of y_{j+1}. All the methods found in popular codes do this in a way that can be viewed as taking each step with two methods of different orders. Note that the approximation of Euler's method is formed in the course of forming the second order Taylor series approximation. This is an extreme case of what is called an "embedded" formula. Generally an embedded formula can be evaluated along with a primary formula at (greatly) reduced cost by taking advantage of computations made in the evaluation of the primary formula. Denoting the solution computed by the (primary) second order Taylor series method by y^*_{j+1}, it is seen that, to leading order, the error of the lower order Euler formula is $y^*_{j+1} - y_{j+1}$. The argument is perfectly general; to leading order the error of a step of order k can be estimated by comparing the result to one of order $k + 1$.

The codes try to use the largest step size that will yield the desired accuracy because this minimizes the number of steps, hence the cost of the integration. Very often adapting the step size to the solution more than pays for the cost of estimating the error. Furthermore, variation of step size is what makes it possible to solve hard problems–a small step size is used where necessary to resolve the solution and a large step size is used wherever possible for efficiency. Automatic adjustment of the step size also keeps the integration from blowing up when solving the "stiff" problems discussed in Section 6. But, the most important reason for estimating the error and insisting that the code deliver a result of at least a specified accuracy at each step is that it gives us some confidence in the numerical solution.

Two techniques are used to get more accuracy. One seen in all production codes is to reduce the step size. The other is to change the number of terms in the Taylor series. Correspondingly, some of the most widely used codes, viz. those based on Adams formulas, the backward differentiation formulas (BDFs) popularized by Gear, and extrapolation of a midpoint rule, vary the order of the method used so as to approximate the solution as efficiently as possible [35, 12].

Although the integration is advanced with the value at x_{j+1}, it is obvious from the derivation of Taylor series methods that they provide an approximate solution throughout the subinterval $[x_j, x_{j+1}]$, which is to say that a continuous extension is obvious. In aggregate these approximations provide a numerical solution throughout $[a, b]$. There is a question about how smoothly the approximation for one subinterval connects to that of adjacent subintervals. Popular codes differ in that they may or may not have continous extensions and when they do, the approximate solution may, or may not, be $PC^1[a, b]$. The design of production codes assumes only that $f \in PC[a, b]$, hence $y \in PC^1[a, b]$, so to deliver more smoothness as a matter of course would be inconsistent. This assumption is made because this is approximately the minimal smoothness that guarantees the existence of a unique local solution and the aim is to make the codes applicable to as wide a class of problems as possible.

5. Is the solution any good?

A code for IVPs does not try to do what a casual user might expect. There are two main ways of describing what it attempts. One is that it attempts to track accurately the local solution at each step. As we saw in the last section, this view of the matter is particularly natural when developing a method. We cannot expect a numerical method to do a perfect job, meaning that generally y_{j+1} is not exactly equal to $u(x_{j+1})$. Accordingly, when we take the next step, we approximate a *different* local solution, one defined by the initial value y_{j+1} at x_{j+1}. Our goal is to approximate *the* solution $y(x)$ of the given IVP, but the code approximates instead local solutions that originate near $y(x)$. All we can expect of the numerical method is that it produce a result "close" to the current local solution at each step. The implications of this follow from the behavior of the solutions of the ODE, a matter that we believe ought to receive emphasis in courses on ODEs rather than the formula for taking a step. If solutions that originate near $y(x)$ spread out, which is to say that the solution of interest is unstable, then even one small error that moves us onto a solution with very different behavior will lead to approximations that are very different later in the integration. Conversely, if $y(x)$ is stable, errors are actually damped out as local solutions approach $y(x)$.

Gronwall's lemma establishes the stability of ODEs that satisfy a Lipschitz condition with constant L. Concretely, if $u(x)$ and $v(x)$ are solutions of such a system, then for $x \geq \xi$,

$$\|u(x) - v(x)\| \leq \|u(\xi) - v(\xi)\| \exp L(x - \xi).$$

The scalar equation $y' = Ly$, $L > 0$, shows this bound to be sharp. Suppose that a code is tracking a solution u and an error of ϵ is made in stepping to ξ. If $v(x)$ is the new local solution with $v(\xi) = u(\xi) + \epsilon$, the inequality bounds the effect of the error at later x. This is a bound on the error that would be present even if *no* error is made at subsequent steps. As the scalar example makes clear, a single error ϵ can result in an error at b as large as $\|\epsilon\| \exp L(b-a)$. Certainly we cannot, in general, expect numerical errors smaller than this. The theory of stable numerical methods shows that when the maximum step size is H, introducing errors that are $O(H^{k+1})$ at each of $O(H^{-1})$ steps will result in errors that are at worst $O(H^k)$. The cumulative effect of the errors is multiplied by a factor of $\exp L(b-a)$, reflecting the stability properties of the ODE. Clearly, if $L(b-a)$ is "large", the approximate solution might be so inaccurate that it is unusable even though the error at every step is "small". That is why a problem in the "classical situation" is defined as one for which $L(b-a)$ is "not large".

The theoretical bound on the growth of numerical error provides some comfort, but it is often not even qualitatively correct in practice because the Lipschitz condition is not sufficiently sharp. For instance, the scalar equation $y' = -Ly$ has the same Lipschitz constant as $y' = Ly$, but all solutions converge exponentially fast rather than diverge as for the latter equation. With an understanding of what codes do, we now appreciate that solving these two problems numerically differs in important ways. In practice it is the anticipated stability of the problem that furnishes guidelines as to what we can hope to achieve. In particular, we cannot expect to compute realistic numerical solutions unless the problem is at least moderately stable. It is natural to expect that the error will accumulate steadily as an integration proceeds, but that is not so–the stability of the problem can cause errors to decrease as well as to increase. Striking examples arise when integrating from a to b and then back to a. There are a number of technical reasons why the direction of integration might matter, but the one important here is that the stability of an IVP can depend strongly on the direction.

The other main way of describing what codes attempt is in terms of a continuous extension of the method. Suppose that we obtain from the code an approximation $S(x) \in C^1[a,b]$ that is uniformly of the same accuracy as the values $y_j = S(x_j)$ at mesh points. (A PC^1 approximation would do, but the additional smoothness simplifies the discussion.) Some recent codes actually provide such an approximation, but to understand what codes accomplish, we need only assume the existence of an $S(x)$, a fact not difficult to prove. The approximation satisfies the differential equation with a residual

$$r(x) = S'(x) - f(x, S(x)).$$

The code attempts to produce a solution $S(x)$ with a "small" residual. Indeed, for a method of fixed order k and a maximum step size H, the residual is uniformly $O(H^k)$. Put differently, the *numerical* solution $S(x)$ is the *analytical* solution of the differential equation

$$v' = f(x, v) + r(x, H)$$

with initial condition $v(a) = A$. How close $S(x)$ is to $y(x)$ is then a standard question in the theory of ODEs about the effect of perturbing the given ODE. In the first instance this is a matter of stability, but a sharper result shows that if the residual $r(x, H)$ is uniformly $O(H^k)$, then $y(x) - S(x)$ is also uniformly $O(H^k)$. In this view, what a code does is to compute an approximation that "almost" satisfies the given ODE. How close this approximation is to the solution of the given problem is then a matter of the stability of the IVP.

It is a familiar phenomenon in perturbation theory that a small, persistent disturbance can have a large effect after a sufficiently long time. Suppose we solve $y'' + y = 0$ with initial values $y(0) = 0, y'(0) = 1$ on an interval $[0, b]$ and produce a numerical solution $S(t)$ such that $S(0) = 0, S'(0) = 1$ and $S''(t) + S(t) = 2\epsilon\cos(t)$ for some "small" ϵ. This is all we can expect of a code, but $S(t) = (1 + \epsilon t)\cos(t)$, so when $b = O(1/\epsilon)$, this numerical solution is useless. If the problem is posed on an interval this long, it is not in the classical situation despite having a Lipschitz constant that seems small. As this example shows, we must be cautious about accepting numerical results computed over "long" time periods.

In either view, the goal of a code is to produce the solution to a problem close to the one given. In the first view, the differential equations are the same and initial conditions are perturbed at each step. In the second, the initial conditions are the same and the differential equations are perturbed in the span of each step. Whether accomplishing this goal results in a solution close to that of the given problem is a question of the stability of the given IVP. This question is outside the terms of reference of the codes. Indeed, it is often beyond the grasp of the person setting the problem, but some insight into the stability of the IVP is crucial to a proper interpretation of numerical results.

6. If a little stability is good, how about a lot?

In the last section we saw that the stability of an IVP is fundamental to its numerical solution. A problem that is *very* stable seems ideal because any error that arises is damped out quickly. Unfortunately that is not so, and here's why: Clearly the stability of the numerical method must imitate that of the IVP itself. Methods that imitate the stability of IVPs for all sufficiently small step sizes are called "stable". Not all plausible numerical methods are stable in this sense, but of course, only stable methods are found in production codes. Fine, but how small is "sufficiently" small? In the classical situation a step size that will yield modest accuracy is small enough. However, it is easy to imagine that in other situations, a step size that would yield the desired accuracy might be very much larger than the step sizes that would yield a stable computation. Such problems are called "stiff". The essential ingredients are that the IVP is very stable and the solution is easy to approximate. Stiff problems are not in the classical situation, so for a problem to be stiff, it is necessary that $L(b-a)$ be "large", but this is *not* sufficient.

To illustrate stiffness, consider the computation of the solution $y(x) = 1 - \exp(-100x)$ of $y' = 100(1-y)$ for $0 \leq x \leq 10$ with $y(0) = 0$. The difference between any two solutions $u(x)$ and $v(x)$ of the ODE is a multiple of $\exp(-100x)$, so the problem is very stable on this interval. Suppose now that u_j, v_j are approximations to $u(x), v(x)$ at $x = x_j$ respectively. In a step of length h, the difference between the numerical solutions computed by Euler's method is

$$\begin{aligned} |u_{j+1} - v_{j+1}| &= |(u_j - 100hu_j) - (v_j - 100hv_j)| \\ &= |1 - 100h||u_j - v_j|. \end{aligned}$$

Obviously it is necessary that $h < 2 \times 10^{-2}$ if the numerical solutions are to imitate the stability of the IVP even in the qualitative sense that they merely become closer. More generally, for Euler's method to imitate the stability of the IVP, the step size must be $O(L^{-1})$. Unfortunately this is characteristic of standard methods. For small h, the error of Euler's method is about $h^2 y''/2$. Here this is $h^2 10^4 \exp(-100x)/2$. The solution has an initial transient where it changes rapidly. For Euler's method the step size h must be rather smaller than 10^{-2} to get any accuracy in the transient, where x is small. Likewise, all standard methods must use a small step size to resolve the transient. The problem is not stiff in the first part of the interval because a step size that will yield the desired accuracy is small enough that the stability of the numerical method imitates that of the IVP. However, the solution tends rapidly to a constant and all standard methods are exact for a constant solution. The error expression for Euler's method makes this obvious, at least to leading order. Although the solution can be approximated very easily once past the transient, it is still necessary to use a step size no greater than 2×10^{-2} to keep the computation stable. This surprising and frustrating phenomenon is what is meant by stiffness. Monitoring of the error and adjustment of the step size will keep a code from blowing up due to instability, but the step sizes necessary for stability of the numerical method seem absurdly small for computing a solution that is very easy to approximate. It is enlightening to consider the (quadrature) problem $y' = 100\exp(-100x)$ with $y(0) = 0$, which shares the solution $y(x) = 1 - \exp(-100x)$. This problem has a Lipschitz constant 0, hence is an example of the classical situation. The stability of the IVP is neutral and the numerical integration of this IVP presents no difficulty once past the transient–for a problem to be stiff it is necessary both that the solution be easy to approximate and that the problem be very stable.

It would be nice if we could ignore stiffness, but we can't; stiff physical problems are too important and their solution is impractical without special methods. Generally stiffness arises when phenomena occur on very different time scales and it is a solution evolving on a relatively slow scale that is of interest, such as is often the case in chemical reaction problems. The difficulty is that standard numerical methods will, in effect, use a step size short enough to resolve the fastest possible change even though the solution of interest is changing at a much slower rate. Numerical analysts have developed methods appropriate for stiff problems, with the backward differentiation formulas (BDFs) being the most widely used. Compared to methods appropriate to the classical situation such as Taylor series, explicit Runge-Kutta, and Adams, these methods are very expensive per step when implemented in a manner necessary for solving stiff problems. However, such a method can be used with a step size that is determined by the accuracy required rather than the stability of the ODE. If there are very different time scales present, such methods allow a step size suitable for resolving phenomena on the scale at which the desired solution evolves, not the fastest scale possible. Each step is expensive with such methods because they gather information about the stability of the problem by means of the Jacobian matrix $\partial f/\partial y$. Forming the Jacobian can be expensive, storing it can be expensive, and solving repeatedly at each step linear systems involving it in an iteration to form y_{j+1} can be expensive. Even in the most favorable case this iteration is significantly more expensive than the corresponding calculations of standard methods. Such an expensive step is worthwhile only if the method permits steps sufficiently longer than those used by a conventional method to amortize the cost.

There is no satisfactory mathematical definition of a stiff problem. The practical definition is that it can be solved very much faster using a code specifically intended for stiff problems than a code for standard problems. Some identify any problem difficult for a code intended for non-stiff problems as stiff, but this is simply not correct. A code for stiff problems can be used for non-stiff problems, but will be comparatively expensive. There are problems that are difficult for both the non-stiff and stiff methods found in popular codes, e.g., problems with many discontinuities or with highly oscillatory solutions. Nowadays, there are codes for non-stiff problems that diagnose if stiffness is the reason for any unsatisfactory, seemingly inefficient performance. Also, there are more experimental codes that implement both stiff and non-stiff solvers and switch between the solvers on the basis of similar diagnostics.

7. From IVPs to BVPs.

Considering their importance in practice, scant attention is given to BVPs in ODE textbooks. More attention to BVPs would illuminate the usual study of IVPs, both with respect to the theory and to the use of numerical methods. Although not so well-known as codes for IVPs, there are quality codes for BVPs that can be obtained easily enough. Here we discuss briefly some issues that might be taken up early in ODE courses.

In many introductory texts, problems are taken up that are actually BVPs, though this may not be stated explicitly. Simple problems can be used to illustrate some basic facts. For instance, it is easy to discuss the existence and uniqueness of solutions of $y'' + y = 0$, $y(0) = 0$, with $y(b) = B$ for some $b > 0$ by considering solutions $y(x, s)$ of the equation with initial values $y(0) = 0, y'(0) = s$. It is found that the boundary condition at b leads to a linear equation for the slope s and that there is no solution, exactly one, or infinitely many, depending on b and B. This is already quite different from IVPs, but the behavior is seen to be still more complex for the nonlinear equation $y'' + |y| = 0$. The same method of analysis can be applied with little difficulty–the original example is treated in detail in [4], pp. 4-10, and as an exercise with hints in [2], p. 127. It is found, e.g., that for $b \geq \pi$, there is no solution for $B > 0$, for $b > \pi$, there is one solution for $B = 0$ and *two* solutions for $B < 0$, and for $b = \pi, B = 0$, there are infinitely many solutions. BVPs arise in more diverse forms than IVPs, so preparing them for popular codes is more complicated. The examples here are called two-point problems because boundary conditions (BCs) are set at the two ends of the interval of interest $[0, b]$.

Codes for two-point BVPs based on shooting (or marching) methods imitate this method of analysis. They guess the unknown solution values at one end of the interval to obtain an IVP, integrate with a quality code for IVPs to the far end of the interval (a process called "shooting" or "marching"), and see if the computed solution satisfies the BCs at that end. Satisfying the BCs amounts to constructing a set of algebraic equations for the unknown initial values. When the ODEs are linear, so are the algebraic equations, and existence and uniqueness is clear in principle. However, when they are not linear, the situation is murky. This makes BVPs harder to solve because, as we have seen, it is perfectly possible that there be no solution. And, if there is more than one solution, there is a question about which to compute. For this reason shooting codes for BVPs ask users to guess the unknown initial values and try to approximate initial values for a solution of the BVP that are nearest to the guessed values. All this makes the numerical solution of BVPs much less routine than IVPs. It can be hard to decide whether a code has failed to converge because the BVP does not have a solution, or it is very "ill-conditioned" or "unstable" because its data is close to that of a BVP with no solution, or the guess for the initial values is not good enough, or the method of the code is inappropriate for this particular BVP.

Shooting methods can be quite effective, but a great many problems simply cannot be solved in this way. The reasons are illuminating. The fundamental difficulty is that a stable BVP often leads to IVPs that are not stable. The situation is exposed by considering problems on $[0, 1]$ of the form $y'' - (\alpha+\beta)y' + \alpha\beta y = 0$. The solutions of the ODE are linear combinations of $\exp(\alpha x)$ and $\exp(\beta x)$. If, say, $\alpha = -1$ and $\beta = +100$, it is not practical to shoot from 0 to 1 because the IVP is terribly unstable. Indeed, a common difficulty with shooting methods is that the integration cannot reach the far boundary because the numerical solution blows up along the way. Nevertheless, a two-point BVP for this equation is quite stable. For example, if the BCs are $y(0) = A$, $y(1) = B$, a little calculation shows that $|\partial y(x)/\partial A| \leq 2e$, and $|\partial y(x)/\partial B|$ satisfies a similar bound. If $\alpha = -1$ and $\beta = -100$, shooting from $x = 0$ to $x = 1$ is stable, but it is still not practical to solve some BVPs for this equation. There is more than one reason for this and we refer to [35], p. 79 ff., for a more complete discussion. One issue that we can understand now is that we cannot compute a solution that behaves like $\exp(-100x)$ in the presence of solutions that behave like $\exp(-x)$ because the IVP is unstable, that is, the computed components corresponding to $\exp(-100x)$ are "swamped" by those corresponding to $\exp(-x)$. We might also have some difficulty computing a solution that behaves like $\exp(-x)$ in the presence of solutions that behave like $\exp(-100x)$ because this IVP is stiff.

To deal with the fact that many stable BVPs lead to unstable IVPs when shooting, it is necessary to resort to "boundary value" methods that approximate the solution over the whole interval with no directional bias. It is understandable that considerable care is needed in the efficient handling of the storage of such an approximate solution and the formation and solution of the large linear algebraic systems involved in improving the approximation by iteration. Nevertheless, the boundary value approach is essential for many problems. The essence of the matter is that the numerical method must be able to imitate the behavior of the mathematical problem. In particular, it must be stable (well-conditioned) when the mathematical problem is stable.

8. Where do I get the tools?

To locate ODE software for general scientific computation, try the Guide to Available Mathematical Software (GAMS) [5] at http://gams.nist.gov. One of the sources of software included in GAMS is so important that it demands separate mention: Netlib [13], an extensive collection of quality numerical programs available at http://www.netlib.org and a number of reflector sites worldwide. (The ODE codes are contained mainly in the chapters ode and odepack and in the ACM TOMS library.) Most of the software listed in GAMS is free, including the SLATEC library [15], but the Guide does point to commercial libraries of numerical software. The two most important are IMSL [25] (distributed by Visual Numerics, Inc.) and NAG [30] (distributed by Numerical Algorithms Group, Inc.). Computing environments like MATLAB, Maple, and Mathematica contain special libraries that we'll discuss more in a moment. Some quality software is available by means of books. Indeed, a couple come with small libraries that include a quality code for IVPs, viz. Kahaner, Moler, and Nash [28] includes SDRIV2 and Forsythe, Malcolm, and Moler [16] includes RKF45. As might be expected, specialized texts on the numerical solution of IVPs such as Hairer, Nørsett, and Wanner [21, 22] and Shampine and Gordon [37] include codes, but even the introductory text on ODEs by Sanchez, Allen, and Kyner [34] includes RKF45. The specialized text [35] provides details about, and sources for, many of the most widely used codes for IVPs.

In the sources cited, quality software can be found for IVPs in both the standard first order form and the special second order form, such as [6], as well as in many forms for BVPs, including Sturm-Liouville problems, such as [3]. Both the codes mentioned here are in the ACM TOMS collection. The book [2] contains prologs of several BVP codes, including the powerful COLNEW, which is available from Netlib and the NAG library. In addition there are codes available for related tasks, such as solving differential-algebraic systems in either IVP or BVP form.

Gear's important code DIFSUB popularized in his book [17] makes available two kinds of methods for IVPs, Adams-Moulton and BDF. It is the predecessor of a line of codes developed at LLNL by Hindmarsh, Byrne, and their associates that are in wide use. Principal among them is LSODE [23] and the more recent VODE [9]. Although there are other codes implementing both Adams and BDF methods that are at least as powerful and user-friendly, VODE is a convenient choice because it is a quality code that makes straightforwardly available both kinds of methods and is readily obtained from Netlib. The only reason you might want to use one of its relatives is that some have special capabilities—e.g., LSODI solves linearly implicit ODEs.

The explicit Runge-Kutta code RKF45 of Watts and Shampine [39] is in very wide use, with instances already noted and others appearing below. Similar codes, DVERK, of Hull, Enright, and Jackson [24], and D02PAF, of Gladwell [19], have also been popular, appearing, e.g, in the IMSL and NAG libraries, respectively. The suite of explicit Runge-Kutta codes RKSUITE [7, 8] incorporates advances in the theory and practice of such methods made since RKF45, D02PAF, and DVERK were written. It is a good choice because it makes available three sets of formulas, a number of new features, a collection of templates illustrating use of the suite, and is readily obtained from chapter ode of Netlib. It is coming into wide use, having, e.g., been added to both the IMSL and NAG libraries.

There are not as many codes in wide use as it might first appear because popular codes have been distributed under different names, sometimes without attribution. A little caution is warranted because the codes may have been translated and/or modified without consulting the authors. The sources we list here are reasonably safe.

Quality software makes assumptions about typical problems, the kind of output desired, and the computing language/hardware to be used. The assumptions in computing environments like MATLAB [40], Maple [11], and Mathematica [41] are different from those of general scientific computation. In particular, the emphasis is on convenience and graphical interpretation of results at the possible expense of efficiency in execution time and storage. Because MATLAB is oriented toward numerical computation, it was relatively easy to modify RKF45 for the environment. The resulting code ode45 has been the main code for ODEs through the current release. Subsequently the MATLAB ODE Suite [38] of solvers was written specifically for the environment. Another tack is to use gateway software from MATLAB to standard numerical libraries, including in particular most of the ODE solvers of the NAG library.

Making the popular codes for general scientific computation available in packages oriented towards symbolic computation like Maple, Mathematica, and Axiom [27] is much more difficult. The default in the code dsolve in Maple is to solve a problem by symbolic means. There is an option to solve it by numerical means using a modification of RKF45, and the most recent release includes other widely used solvers such as LSODE. Among the difficulties in such an environment is that the very nature of an IVP can be different. For example, it would be common to have an initial condition that is a variable in algebraic solution, but it must be a number in numerical solution. Further, the nature of the result is different. In the one case a formula for the solution is produced and in the other, what amounts to a table of solution values. In this connection we note that the underlying code RKF45 did not come with a continuous extension, though it is known now how to obtain a good one. We are not familiar with the modifications made to RKF45 for Maple, so do not know how this issue was resolved. The code NDsolve in Mathematica is based on a variant of LSODE due to Petzold and Hindmarsh called LSODA [32] that tries to select automatically Adams or BDF methods. The scanty documentation states that the output is in the form of an InterpolatingFunction object. InterCall [26] is a package that permits Mathematica to access the NAG and IMSL libraries, hence providing access to a wide range of ODE solvers. Axiom does not provide numerical ODE solvers within the symbolic framework, relying instead on a custom-designed bridge to the NAG Foundation library.

Special codes have been written for dedicated use in teaching packages. Naturally they exploit the particular class of problems and the display possibilities of the packages. Typically the kinds of problems allowed are extremely limited, facilitating greatly the solution of the IVP and display of the results. For a review of the major possibilities, including the environments already mentioned, see Flint and Wood [14] or http://www.math.hmc.edu/codee/solvers.html. There is a trend towards bringing into this context the general-purpose codes aimed at general scientific computation. As with the environments, this is not entirely straightforward. For example, the computational engine of ODE Architect [31] is an interface to RKSUITE and VODE. A complication is that the two underlying codes differ in important ways, e.g., the form of the error control. In the ODE Architect context, what constitutes a "typical" problem and a need for simplicity suggests allowing only one of the sets of formulas of RKSUITE. Matching the way that the codes produce output with the display requirements is not easy, especially in view of an animation facility. Clearly the codes in wide use for scientific computation cannot simply be "plugged into" such packages because they were designed for purposes that are rather different. In passing we observe that RKSUITE and VODE are also the two packages chosen for the ODE chapter of the electronic book prepared for the Computational Science and Engineering Project [1], which introduces students to techniques useful in scientific computation. This application is closer to the intended use of these codes, so all their options are available to the student.

The first teaching packages were written for machines with limited capabilities, which by itself was enough to make impossible merely inserting a general-purpose solver. Computing speed and memory have increased so explosively that this is no longer a major constraint. However, calculators are still limited, though they, too, have seen an explosive growth in capabilities. On the TI-85 calculator [29] and its successor the TI-92, the available memory is constrained and the visual display rather limited (particularly on the TI-85), so in designing the ODE solver it was determined to use a method designed for low accuracy calculations with low memory requirements, specifically the (2,3) pair of RKSUITE. Accordingly, the hardware constraints limit the efficiency of solution for the moderate accuracies that might be requested when using these calculators. On the other hand, most of the options of RKSUITE are available via keystrokes.

References

[1] R.C. Allen Jr. et al, Computational Science Education Project, 1996. Electronic book available from `http://csep1.phy.ornl.gov/CSEP/BOOK.PS` and from several reflector sites.

[2] U.M. Ascher, R.M.M. Mattheij, and R.D. Russell, *Numerical Solution of Boundary Value Problems for Ordinary Differential Equations*, Prentice-Hall, Englewood Cliffs, NJ, 1988.

[3] P.B. Bailey, M.K. Gordon, and L.F. Shampine, Automatic Solution of the Sturm-Liouville problem, *ACM Trans. on Math. Software* **4** (1978) 193–208.

[4] P.B. Bailey, L.F. Shampine, and P.E. Waltman, *Nonlinear Two Point Boundary Value Problems*, Academic, New York, 1968.

[5] R.F. Boisvert, S.E. Howe and D.K. Kahaner, GAMS: A Framework for the Management of Scientific Software, *ACM Trans. on Math. Software* **11** (1985) 313–355. GAMS available electronically from `http://gams.nist.gov` and reflector sites.

[6] R.W. Brankin, J.R. Dormand, I. Gladwell, P. Prince, and W.L. Seward, ALGORITHM 670: A Runge-Kutta-Nyström Code, *ACM Trans. Math. Softw.* **15** (1989) 31–40.

[7] R.W. Brankin, I. Gladwell, and L.F. Shampine, RKSUITE: a Suite of Runge-Kutta Codes for the Initial Value Problem for ODEs, Softreport 91-1, Math. Dept., SMU, Dallas, 1991. RKSUITE available from chapter ode of Netlib.

[8] R.W. Brankin, I. Gladwell and L.F. Shampine, RKSUITE: A Suite of Explicit Runge-Kutta Codes, in " Contributions to Numerical Mathematics" (ed. R.P. Agarwal), *World Sci. Ser. Appl. Anal.* **2** (1993) 41–53.

[9] P.N. Brown, G.D. Byrne, and A.C. Hindmarsh, VODE: a Variable-Coefficient ODE Solver, *SIAM J. Sci. Stat. Comput.* **10** (1989) 1038–1051.

[10] J.L. Buchanan, MDEP Midshipman Differential Equations Program, Version 2.29, Math. Dept, U.S. Naval Academy, Annapolis, MD, 1992.

[11] B.W. Char, K.O. Geddes. G.H. Gonnet, B.L. Leong, M.B. Monagon and S.M. Watt *The Maple V Language Reference Manual*, Springer-Verlag, New York, 1991. Maple V is distributed by Waterloo Maple Software.

[12] P. Deuflhard, Recent Progress in Extrapolation Methods for Ordinary Differential Equations, *SIAM Review* **27** (1985) 505–535.

[13] J.J Dongarra and E. Grosse, Distribution of Mathematical Software via Electronic Means, *Comm. ACM* **30** (1987) 403–407. See also quarterly column on Netlib in SIAM News. Netlib available electronically from `http://www.netlib.org` and from many reflector sites.

[14] A. Flint and R. Wood, ODE Solvers in the Classroom, *The College Mathematics Journal* **25** (1994) 458–461.

[15] K.W. Fong, T.H. Jefferson, T. Suyehiro and L. Walton, *Guide to the Slatec Common Mathematical Library*. Guide available electronically from Netlib, 1993; SLATEC CMLIB 4.1 distributed on Netlib and by the National Energy Software Center.

[16] G.E. Forsythe, M.A. Malcolm, and C.B. Moler *Computer Methods for Mathematical Computations*, Prentice-Hall, Englewood Cliffs, NJ, 1977.

[17] C.W. Gear, *Numerical Initial Value Problems in Ordinary Differential Equations*, Prentice-Hall, Englewood Cliffs, NJ, 1971.

[18] I. Gladwell, Shooting Methods for Boundary Value Problems, Chapter 16 in G. Hall and J.M. Watt, eds., *Modern Numerical Methods for Ordinary Differential Equations*, Clarendon Press, Oxford, 1976.

[19] I. Gladwell, Initial Value Routines in the NAG Library, *ACM Trans. Math. Softw.*, **5** (1979) 368–400.

[20] I. Gladwell, The Development of the Boundary-Value Codes in the Ordinary Differential Equations Chapter of the NAG Library, in B. Childs et al (eds) "Codes for Boundary Value Problems", *Springer Lecture Notes in Comp. Sc.* **76** (1979) 122–143.

[21] E. Hairer, S.P. Nørsett, and G. Wanner, *Solving Ordinary Differential Equations I, Nonstiff Problems*, Springer-Verlag, Berlin, 1987.

[22] E. Hairer and G. Wanner, *Solving Ordinary Differential Equations II, Stiff and Differential-Algebraic Problems*, Springer-Verlag, Berlin, 1991.

[23] A.C. Hindmarsh, LSODE and LSODI, Two New Initial Value Ordinary Differential Equations Solvers, *SIGNUM Newsletter* **15** (1980) 10–11.

[24] T.E. Hull, W.H. Enright and K.R. Jackson, User's Guide to DVERK—a Subroutine for Solving Nonstiff ODEs, Rept 100, Dept. of Computer Science, U. of Toronto, Canada, 1975.

[25] IMSL MATH Library 3.0 Manual, Visual Numerics Inc, 1995. The IMSL Math Library is distributed by Visual Numerics, Inc.

[26] The InterCall Mathematica link is distributed by Analytica International Pty, Ltd.

[27] R.D. Jenks and R.S. Sutor *Axiom: The Scientific Computation System*, Springer-Verlag, New York, 1992. Axiom is distributed by the Numerical Algorithms Group, Inc.

[28] D. Kahaner, C.B. Moler, and S. Nash, *Numerical Methods and Software*, Prentice-Hall, Englewood Cliffs, NJ, 1989.

[29] E.P. Merkes, The TI-85 Graphics Calculator Tutorial 1995. Electronic book available from `http://math.uc.edu/~brycw/preprint/ti85/ti85.htm`; TI-85 calculator widely available from electronics and office warehouses.

[30] Numerical Algorithms Group Fortran 77 Library Manual, Mark 17, NAG, Oxford, 1996. The NAG Fortran 77 library is distributed by the Numerical Algorithms Group, Inc.

[31] ODE Architect, Consortium for Ordinary Differential Equations Experiments, Wiley, New York, to appear.

[32] L.R. Petzold, Automatic Selection of Methods for Solving Stiff and Non-stiff Systems of Ordinary Differential Equations, *SIAM J. Sci. Stat. Comp.* **4**(1983) 136–148.

[33] J.C. Polking, *MATLAB Manual for Ordinary Differential Equations*, Prentice-Hall, Englewood Cliffs, NJ, 1995.

[34] D.A. Sanchez, R.C. Allen, Jr., and W.T. Kyner, *Differential Equations*, 2nd ed., Addison-Wesley, New York, 1988.

[35] L.F. Shampine, *Numerical Solution of Ordinary Differential Equations*, Chapman & Hall, New York, 1994.

[36] L.F. Shampine and I. Gladwell, Software Based on Explicit RK Formulas, *Appl. Numer. Math.*, **5** (1996) 293–308.

[37] L.F. Shampine and M.K. Gordon, *Computer Solution of Ordinary Differential Equations: the Initial Value Problem*, Freeman, San Francisco, 1975.

[38] L.F. Shampine and M.W. Reichelt, The MATLAB ODE Suite, *SIAM J. Sci. Comp.*, to appear.

[39] L.F. Shampine and H.A. Watts, Practical Solution of Ordinary Differential Equations by Runge-Kutta Methods, Rept. SAND 76-0585, Sandia National Laboratories, Albuquerque, NM, 1976, and The Art of Writing a Runge-Kutta Code, Part I, pp. 257–275 in J.R. Rice (ed.) *Mathematical Software III*, Academic, New York, 1977, and The Art of Writing a Runge-Kutta Code, II, *Appl. Math. Comp.* **5** (1979) 93–121.

[40] K. Sigmon, *A MATLAB Primer*, CRC Press, 1994. MATLAB is distributed by The MathWorks, Inc.

[41] S. Wolfram, *The Mathematica Book* (3rd ed.), Cambridge University Press, New York, 1996. Mathematica is distributed by Wolfram Research, Inc.

Technology in Differential Equations Courses: My Experiences, Student Reactions

Beverly H. West
Cornell University

I have several reasons for using technology in teaching differential equations:

1. Visualization opens an important door to the concepts and their connections to real world phenomena.
2. Interactive computer graphics empowers us to deal with the majority of differential equations that can *not* be solved with formulas.
3. I am able, far better than in the traditional course, to get students to *think* and have real ownership of the ideas they have learned.

What have we learned from the revolution in teaching differential equations? I have been caught up in it since it began in the early 80's, so I will explain my continually evolving experience, and then try to tell what we have learned (so far) from it. Most important is for instructors to do what they are comfortable with and excited about. There are many different possibilities; you should start simply enough so that success will carry you forward.

Some History

Since 1972 I have been teaching differential equations at Cornell, primarily to biology and social science majors. These nonmathematically-oriented students become very interested in differential equations when they see where such modeling might be of use. I have seen the technology of interactive computer graphics vastly increase these students' exposure, understanding, and competence.

From the beginning, a major component of my teaching in that course has been qualitative solution sketching, which we did by hand using isoclines and making refinements with second derivatives. See [4]. A picture is indeed worth a thousand words—when you can quickly see at once "all" the solutions to a differential equation, you can grasp the overall behavior of solutions and zero in on features relevant to the problem at hand. Of course, this is especially important when no analytic (formula) solution exists. Nonlinear systems, with the graphics technology, are handled as easily as linear systems, which are usually the only ones that can be solved analytically.

About 1980 John Hubbard came to Cornell and taught me about fences and funnels, bringing solid mathematical proofs to back up the qualitative techniques. In 1983 we began to collaborate on a text ([10] and [11]), and that summer when the Macintosh arrived with its inexpensive and easily interactive graphics, we immediately started (with student programmers, in Lisa Pascal, for 128K Macs) to develop appropriate software. The resulting *MacMath* is a dynamical systems package with twelve simple programs in which the student need only type in the equations, then point-and-click to choose an initial condition and see the solution evolve. See [12].

MacMath was popular, and the programs were so small and simple that we could simply distribute disks and send students off to do homework independently, wherever they could find a Mac. An extended upgrade with a Windows version is in production. *MacMath* remains my program of choice, especially for its bifurcation and 3D features.

In 1989 we had at Cornell an NSF grant under the USME program to work on "Enhancement Options for Second Year Calculus Courses". Our visiting Danish colleague Bjorn Felsager developed a full-blown once-a-week laboratory program that has continued to the present. See [9]. For each week, Felsager created a 50 minute lab for students (with partners) to explore a topic, then he provided a choice of about six homework exercises, most often from different real-world applications of the topic at hand, with an instructor's version giving brief solutions including the graphs. This was a successful effort that I have continued to use and extend.

Example A: Felsager's introductory lab D1 introduces students to ODE software with a study of the behavior of $y' = y \cos t$, comparing results of the analytic solution versus different numerical methods and different step-

Matching Exercise:

(a) Match each trajectory in the 3D xyt view with the appropriate solutions in the xt and yt views.

(b) Find the initial conditions (at t=0) as best you can estimate from the 2D views for each trajectory. State them numerically, and mark each on the appropriate 3D orbit.

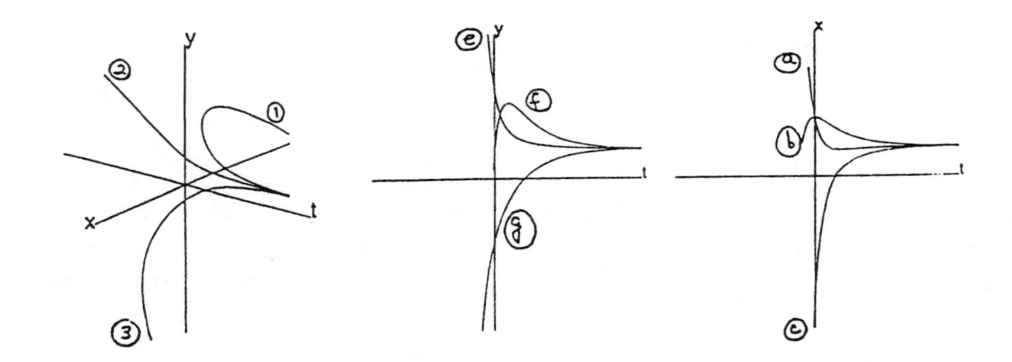

Figure 1: A *MacMath* printout relating different representations; each axis runs from -5 to 5.

sizes. (See Figure 2.) The Homework exercises provide a choice of problems ripe for similar studies.

Before the time of this NSF grant, our classroom use of technology was scant; given Cornell's difficult terrain and weather and the fact that our classes are scheduled all over campus, an occasional demonstration was a major achievement. But from 1989, with the cheerful and competent assistance of Cornell's Academic Computing Services, we were able to use once a week a superb laboratory facility with 40 MacII's that could serve two students each, in place of a regularly scheduled lecture or recitation. Eventually the Mathematics Department created its own smaller but well-equipped Macintosh lab, with additional software; its greater convenience has encouraged more of our colleagues to become involved.

In 1990 Bob Borrelli and Courtney Coleman (see [2], [3], [4]) of Harvey Mudd College secured funding for an NSF Consortium for Ordinary Differential Equations with Experiments (CODEE) that with seven diverse institutions has provided national leadership for the new point of view. See [5], [6], and Table 2. In 1992 we and Anne Noonburg at Cornell and Borrelli/Coleman at Harvey Mudd gave the first CODEE summer workshops for colleagues and were inspired by the enthusiasm and creativity of our participants.

In 1994 I met Hubert Hohn, a talented teacher and graphic artist who had created a satellite television course called *Order in Chaos* involving many interactive illustrations of differential equations and dynamical systems. Hohn's work filled a gap I felt in my students' experience, that of connecting graphs with reality and, by visualizing, bringing a less abstract alternative approach to concepts. Addison Wesley Interactive assembled a team of mathematicians who had collaborated with Hohn and with each other—John Cantwell of St. Louis University, Jean Marie McDill of Cal Poly San Luis Obispo, Steven Strogatz and myself from Cornell—to publish *Interactive Differential*

Table 1: Homework exercises (abstracted) from Felsager's Lab D1

Students are instructed to choose any *one* of the following exercises; #3–7 include a paragraph or two of background material and then ask several questions in the context of the actual modeling situation.

Exercise 1. By a study similar to the lab procedure, make an analysis of the equation $y' = (y-1)\cos t$ for $y(0) = 2$.

Exercise 2. By a study similar to the lab procedure, make an analysis of the equation $y' = 2ty - 2t + 2y - 2$ for $y(0) = 2$.

Exercise 3. Exponential growth. For $y' = ky$, compare the doubling times for $k = 1\%, 2\%, 4\%, 5\%, 10\%, 20\%$. Conjecture a rule of thumb.

Exercise 4. Exponential decay combined with linear growth. Mercury poisoning in Minamata, Japan in the early 50's, $y' = a - by$.

Exercise 5. Logistic growth, the Verhulst model. $y' = r(1 - y/k)y$.

Exercise 6. Chemical reactions. $a' = -ka(a - (a_0 - b_0))$, for types A and B molecules dissolving in a liquid reaction to produce molecules of type C.

Exercise 7. Super-exponential growth. $y' = ay^2$; $y' = ay^2 - 1$.

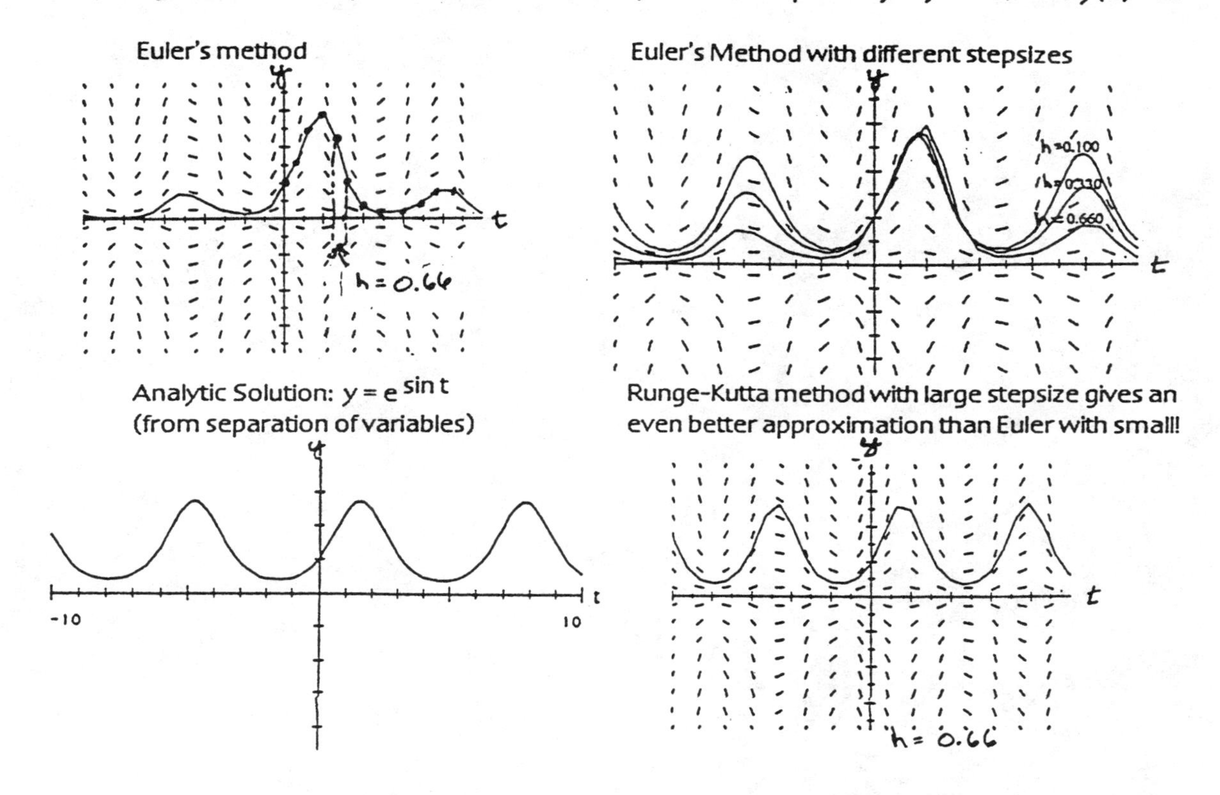

Figure 2: From the instructor's manual for Felsager's Lab D1.

Equations (IDEs), and the next chapter of our saga was underway. See [13] and Figure 3.

Meanwhile, the NSF has funded CODEE for a new venture, in multimedia. We are currently completing *ODE Architect,* a CD-ROM with workbook that will provide a complete laboratory package, based on the combined experience of the Consortium. See [14] and Figure 4.

Helping to coordinate and edit this project has brought me to Harvey Mudd College for a year, where I have had the engaging opportunity to teach an entire semester of differential equations to students who are generally closer to mathematics than my biological and social science students at Cornell. Therefore I am able to give a more balanced perspective and survey. I am grateful to those who have made this possible.

I am more convinced than ever that the visualization and understanding produced by interactive computer graphics greatly enhances and extends a differential equations course. Other authors in this volume do well at elaborating on this point. Instead I stress that differential equations software can be used in a variety of ways—in the classroom or lab, or for homework; subsequent sections discuss this in some detail.

Although I personally favor *MacMath* as an open-ended solver and IDE for concepts, most software coming my way gets used in all of the above ways. Many solvers now exist, and each has some advantages. I am happy for my students to use for homework whichever they like best. I feel it is important for students to have control over such choices, and the resulting variety enhances the homework process—different approaches and graphics serve as a good springboard for discussions that can truly involve students. It is one way to get them *thinking,* which is my ultimate goal and measure of success.

What have I learned?

Present software in a variety of ways

Computer software is a flexible resource that you can use to enhance a variety of teaching strategies and course descriptions.

- A short classroom demonstration can shortcut a lot of lecture time. But be careful to keep it to just a few minutes, or you may not be able to keep your audience awake.

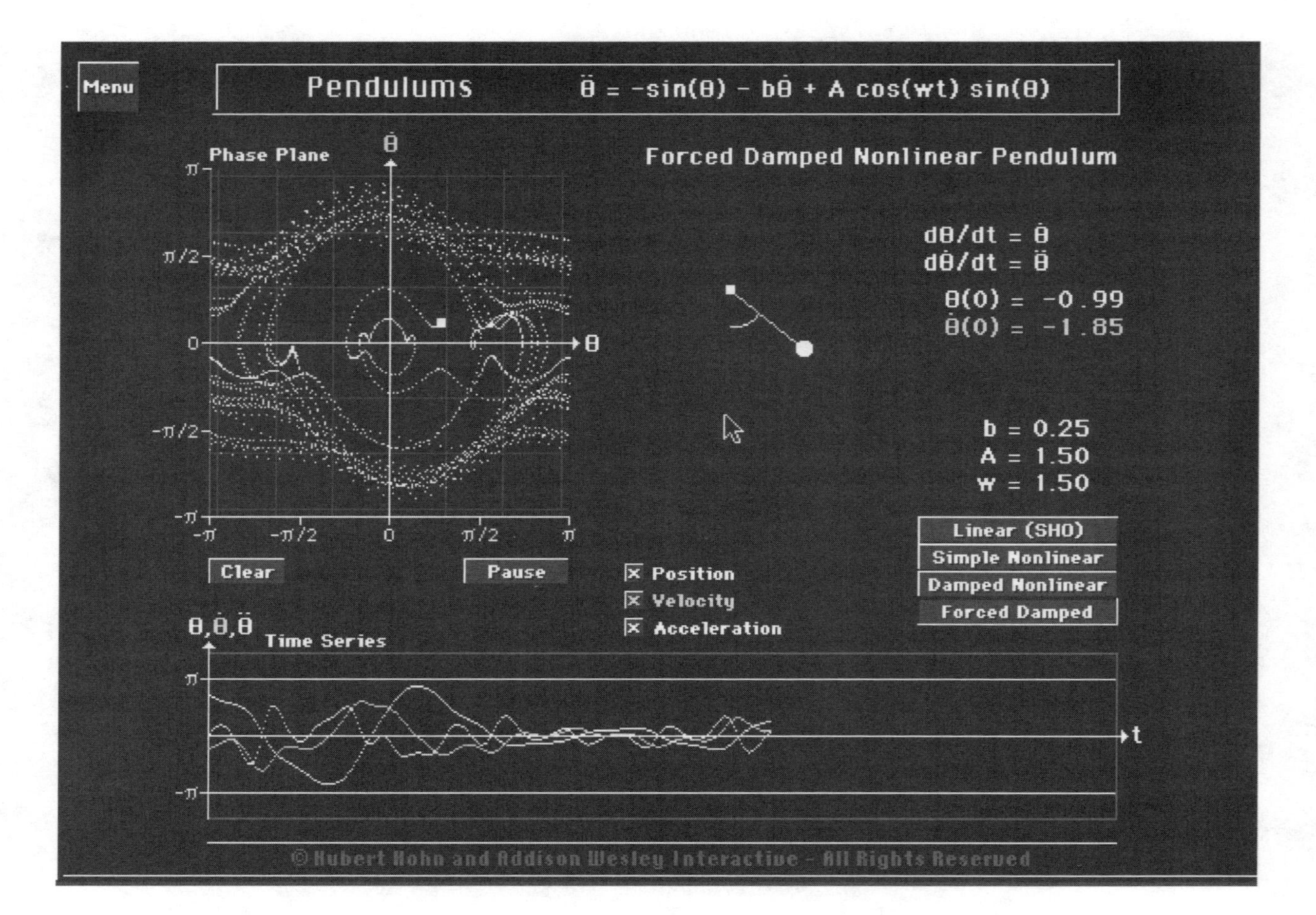

Figure 3: *IDE* shows the motions of a forced pendulum, linked with evolving color-coded graphs of θ, θ' and θ'' as functions of t, and a "phase portrait" of θ and θ' plotted as parametric functions of t.

- You may have the luxury of full-blown laboratory periods to explore a topic in detail. Many resources are available, so you can avoid time-consuming preparation.

- Students can work independently on homework involving computing.

- Students also can work in groups on labs, workbook exercises, or class projects. The interaction provides enhanced learning opportunities, and fewer papers cuts down on grading, so both students and instructors win in this option.

- Minimize frustrations—give a little orientation, anticipate trouble spots. For example, printing has often been a problem, so I usually say "Sketch or print..."—the very act of sketching sharpens the student's focus on what is really important.

Provide choices

Giving choice is an important ingredient for keeping students excited, and amazingly enough, it is not so hard! Felsager started it, and I have continued to extend and use these sets year after year—every term almost every exercise is chosen by some students, which we feel is further "proof" that choice works to maximize students' sense of involvement. They also learn a lot indirectly just by reading the problems from which to choose.

I had feared that making such sets with so many choices would be very difficult, but it actually turns out to be easier than making an assignment that can fit all students—and the individual problems can be much harder. Some students tell me they deliberately choose the one they think will be the hardest.

With solution sheets provided to graders, and with students working in pairs, homework or lab grading is not difficult. Such solutions exist with our published labs, otherwise I either provide them myself, or use solutions

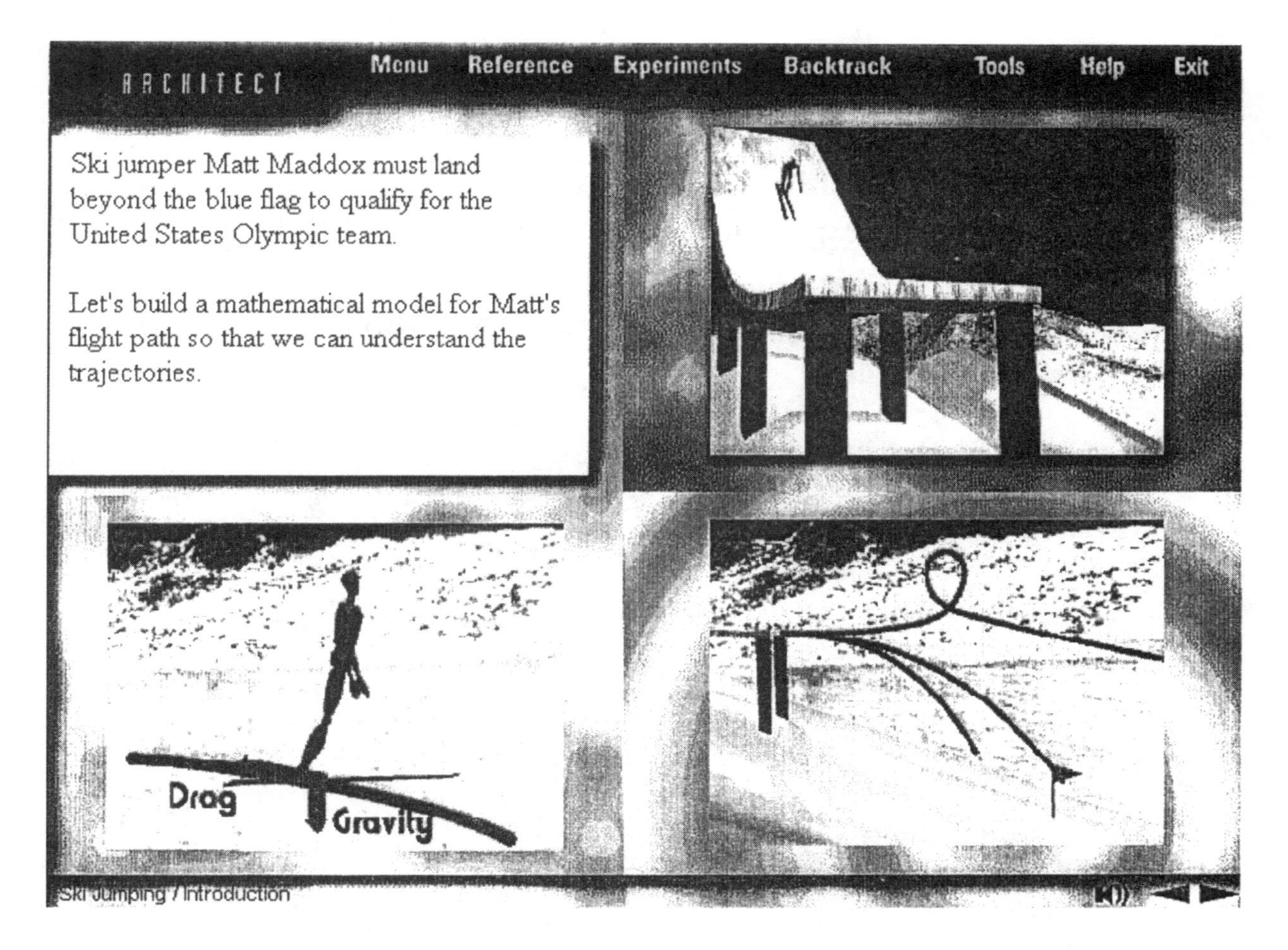

Figure 4: Still shots from the skijumper animation from *ODE Architect*.

that students provide, so that each problem I am grading has a single page solution for reference. The result is that grading goes quickly, but is not boring.

Share answers

I'm always on the lookout for creative approaches and results, and give these attention—either individually or by sharing key insights with the class. Sometimes it is I who does such sharing (always with attribution), by a note or picture on the assignment sheet or on a foamboard poster that I can carry easily to another classroom, and sometimes I ask the students to take a few minutes to show their work.

Such sharing provides exposure to a wider variety of examples than an individual might work on. This also gives the students the feeling their work has some importance, and the quality rises as they see more possibilities. Attaching value to the student's own thinking and expression produces far more careful and sustained thought.

We take care to post both handwritten and fancier computer versions to stress that a nice presentation can be made in either medium. (We don't want students spending extra time creating computer versions unless that is their preferred mode.)

Invite student projects and presentations

One of the simplest ways to do this is to ask for the final lab assignment, on a choice of three topics, to be written up as a presentation, and give the students two weeks to work on it. The last class period or two can then be devoted to brief oral presentations on each topic chosen. We allot about 15 minutes for each topic, most of which have been approached by two or three teams, and ask each team to prepare roughly five minutes worth. We choose a team to begin each presentation, and the others work out how to add do it. The audience actively participates, asking excellent questions and making constructive suggestions. The students do a fantastic job, and it is a

Table 2: The CODEE Consortium 1997

C-ODE-E
Consortium for Teaching Ordinary Differential Equations with Experiments

National Science Foundation

Harvey Mudd College
Robert Borrelli, Courtney Coleman, Michael Moody

Cornell University
Beverly West, John Hubbard

Rensselaer Polytechnic Institute
William Boyce, William Siegmann

St. Olaf College
Arnold Ostebee

Stetson University
Margie Hale, Michael Branton

Washington State University
Michael Kallaher, Thomas LoFaro, Kevin Cooper

West Valley Community College
Wade Ellis, Douglas Campbell

Associate Members
Larry Shampine, Southern Methodist University
Anne Noonburg and Ben Pollina, University of Hartford

terrific grand finale. It gives them a real sense of accomplishment from their semester of study, and a peek at far more than we could have "taught" in a traditional course.

Again and again, it is during final presentations that I observe some of the "weakest" students (according to test scores) being the first to figure out the explanation for some anomaly that has been questioned. See Examples B and C below.

Focus on writing

Although it is well-known (e.g., David Smith with the Duke Calculus Project) that writing in mathematics classes is a very good thing, we never seemed to have time to do very much of it until we came across a wonderful little book by Joan Countryman, *Writing to Learn Mathematics* [7], full of reassurance that it could be done low key. The results were beyond our wildest expectations!

Describing their observations in writing leads students to become far better at expressing mathematical ideas. For example, I often ask students to write for just five minutes in class on Fridays about the topic of the week, what it was good for, and where they were having difficulty. These informal essays are fascinating to read and quickly clue us in to points that needed further clarification. No "extra" time is required of students, and the reading of them goes very quickly. Grading is simple—most papers receive Satisfactory, a few get Excellent, and only one or two papers all term need be marked Unsatisfactory. Nevertheless, the improvement on communication in reports and computer homework has been marked, especially in combination with the "sharing" described above.

This writing also has led to a quantum jump in student effort and success at final exam time, when I have often given a choice of two essay questions they can prepare ahead and consult with me about, but they must write the short essay during the exam. One example asks for a broad perspective on linearization as a recurring theme of differential equations and multivariable calculus; another asks for a description and comparison between analysis of phase portraits in differential equations and analysis of extrema in multivariable calculus. In the semesters with the five minute essays on Fridays, students have far more perspective than in previous semesters, and they have prepared superbly well-organized, insightful, and comprehensive (but succinct) exam essays.

Ask new questions

Yes, the use of computer graphics indeed changes the questions we can ask, and one of the best things about it is that it allows students to approach them in different ways.

One of my favorite types is the matching exercises (see Figure 5), introduced to me by Michèle Artigue and Véronique Gautheron of the University of Paris, who wrote, with an early HP plotter, the first book on studying differential equations from their pictures [1]. I prefer these as they were given to me, simply equations and pictures, to be matched up in whatever ways make the most sense to the individuals involved—best is when students disagree on some answers and argue between themselves to settle which is correct. Whenever their minds are actively engaged like this, I know they are actually learning something, and that it will last. This has been ever so much more effective than the old days when I tried to tell them how to avoid pitfalls.

Determine which of the numbered graphs represents the family of solutions to each of the following equations. More than one graph may correspond to one equation, or a graph may not correspond to any of the equations. In the latter case, try to guess what the differential equation might be, and check your guess with a graphical ODE solver.

(a) $y' = \sin ty$

(b) $y' = y^2 - 1$

(c) $y' = 2t + y$

(d) $y' = (\sin t)(\sin y)$

(e) $y' = y/t^2 - 1)$

(f) $y' = (\sin 3t)/(1-t^2)$

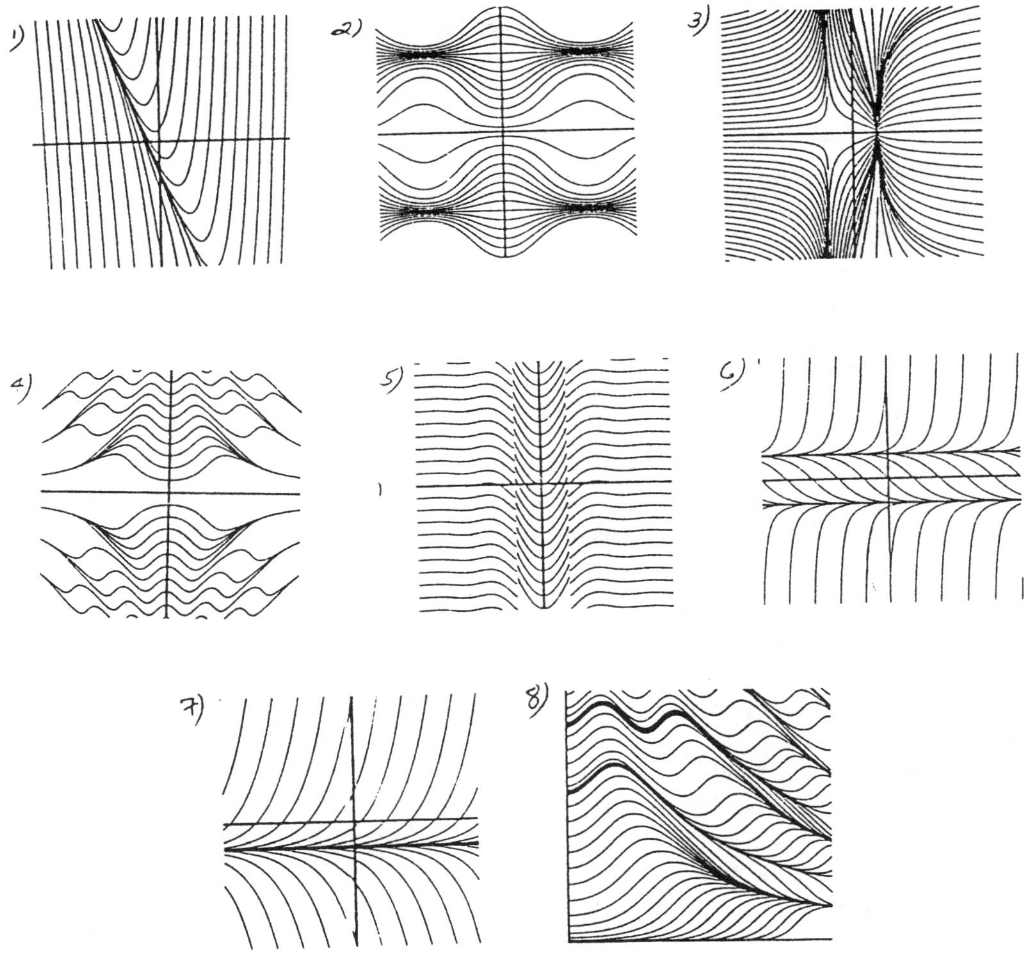

Figure 5: Matching exercise by Artigue and Gautheron

Relax—you don't need to have all the answers

In fact, it is better if you don't, because then the students can continue to engage their minds while thinking on the problem, and many of them can be solved this way. Often I will know an answer but am able to say "you're on the right track—keep going and I think you'll get it figured out".

If I in fact do not know an answer I'll suggest we both work on it "for next time"—only rarely do they fail to get an answer on their own, and if they do fail, I have had time to figure it out with my own resources and colleagues. See Examples B and C below.

Student Reactions

Almost all students "grade" almost all of our weekly laboratory assignments with an A or a B. Interpretive questions are popular—they tend to be seen as more interesting. Students seem the happiest, and most excited, when they have an open-ended lab that allows them to develop and manipulate their own system.

I was surprised to find that although I often use a demo to introduce a topic and expect the lab to "teach" it, many students claim to get the major benefit at the end of a week by using IDE as a summary. This is probably an excellent example of different strokes for different folks.

In general, the students like working with partners; this strategy is seen less as a novelty today than just a few years ago. We have been flexible about who works with whom, but we keep watch and sometimes regroup teams that exhibit some problems to improve morale. We've used a variety of techniques to be sure everyone contributes, and usually they do.

On the negative side, students have little tolerance for frustrations, which we try to minimize but cannot seem to eliminate. They resent feeling the computer work as extra, so it is always a challenge to incorporate technology assignments as a natural part of course. Some actively dislike the workbook concept because tearing out pages to hand in feels like first grade. And they don't want "busy" work like sketching too many graphs, or telling them nothing new.

In fact, the workbook idea poses another pedagogical danger, the fact that students can feel "finished" just because they have filled out a lab sheet, without capturing the real understanding instructors seek. Naturally, the very same questions that are labelled "mindless" by some students are embraced by others as "really helping to clarify concepts"—it is not possible to satisfy everybody at once, which is why variety and choice are so important.

Of course the fact that I groove on pictures doesn't mean that everyone will. It is not uncommon to have some student who just doesn't like labs, or computers, or having a partner; in such a case we can often work something out, but have not found any magic answer.

Specific Examples of Student Interaction:

Example B: During recent years I have been most successful assigning a choice of three topics for final presentations, so that the class can all see each of them on the last day. I'll typically choose one team to lead off with topic A and then expect the others who had worked on that project to add to it; but what has actually happened more often is that more questions than additions arise, and the students, not the instructor, work out the answers. My favorite example was the time that several teams had worked on a combat model of a guerilla force against a traditional force. The overhead projector had failed, so the students suddenly were reduced to having one team member sketch the results while another was explaining (and this has been such a good device that I have never returned to the overhead). Anyway, the team drew some graphs of the guerilla population versus time and the conventional force versus time, as well as a phase portrait for the given initial condition.

One student noted that something was wrong—the guerilla graph began to rise, even though these troops had no reinforcements. Much murmuring ensued as everyone agreed that was crazy. One group turned to me for the answer, but my head was busy with something else, so they continued to argue. Suddenly a student (who had a C-average) observed that the problem was over before the guerilla graph would get that far, because the conventional troops had already been eliminated by that time, and henceforth, in the model, would be negative in number! The students noted that this provided a nice illustration of where a mathematical model reaches its limitations.

Example C: When I was asked suddenly to substitute for a colleague who had not once taken his students into the lab, I asked only for permission to do so. They were studying linear second order differential equations with constant coefficients, so I hastily made a list of four second order equations from four different phase portraits sitting on my desk.

Conventional vs. Guerilla Combat, Lab D8 Exercise

The Combat Model:

conventional troops:	$x' = 0.2 - 0.9y$	$x(0) = 5$ (in thousands)
guerilla troops:	$y' = -0.4xy$	$y(0) = 8$ (in hundreds)

Who do you think will win?

The axes in the following graphs all run from -5 to $+8$.

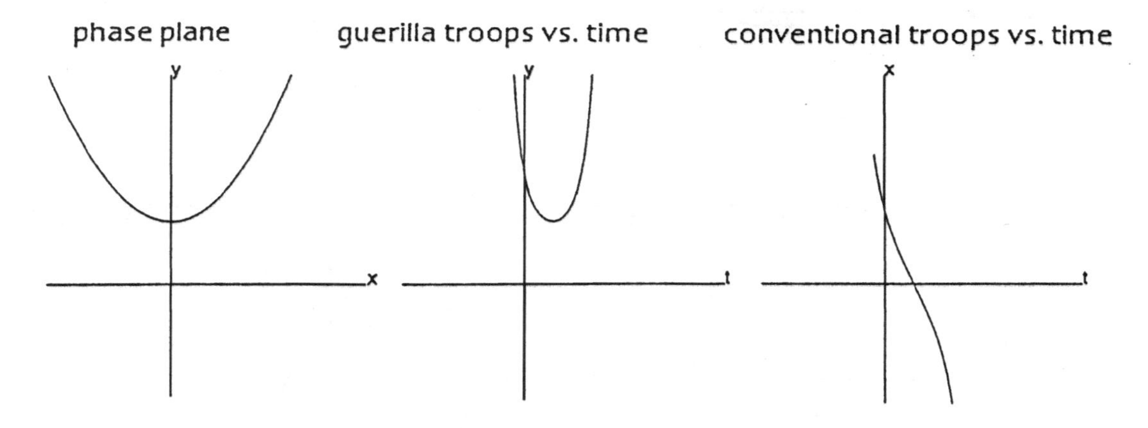

Figure 6: Question: How can the guerilla population rise (without reinforcements)?

I asked the students to write each second order equation as a system of two first order equations, starting with $x' = y$, then to also write it as an equivalent matrix equation. At the computer they were asked to enter the systems into *MacMath* and print out typical resulting phase portraits.

Then they were to find the eigenvalues and eigenvectors of the coefficient matrices—some used *MacMath*, a couple used *Mathematica*. Even though they could of course (at least in theory) have found them by hand, the point was to compare these four examples in one 50-minute lab period, their first.

Lastly the students were asked to draw the eigenvectors onto their phase portraits. Amazed, delighted gasps went up as they did so—these were juniors and seniors in a second course in differential equations, but they had never seen this connection of differential equations with linear algebra. (I hope that phenomenon will soon be fading from the American scene.)

The role of the eigenvectors, and of the sign of the eigenvalues, is very obvious in the saddle and node; in the degenerate case where one of the eigenvalues is zero, I especially liked the student observation that "if one of the eigenvalues is zero, then there is nothing pushing the trajectories to the side as they go along the direction of the other eigenvector". But to the question I always ask for that same picture, "How come all the trajectories stop at the horizontal axis?", no student has ever responded with the answer I am looking for: "Uniqueness of solutions" (because the horizontal axis is either a solution itself or a whole line of equilibrium points).

The spiral phase portrait presented a dilemma—how even to graph eigenvectors that were complex? This is a question I sent everyone away to contemplate, and next lecture we discussed the fact that if you graph real and imaginary parts you will get two vectors that align with the axes of the ellipse the phase portrait would show if there were no damping. It was a good challenge to us all to be puzzling about this overnight, and involved more students than the ones who had tried to come up with an answer in class.

Conclusion

Use technology, in different ways. Avoid overloading yourself, but offer students ways to do their own thing.

The result will bring excitement to the differential equations course, both for the students and the instructors.

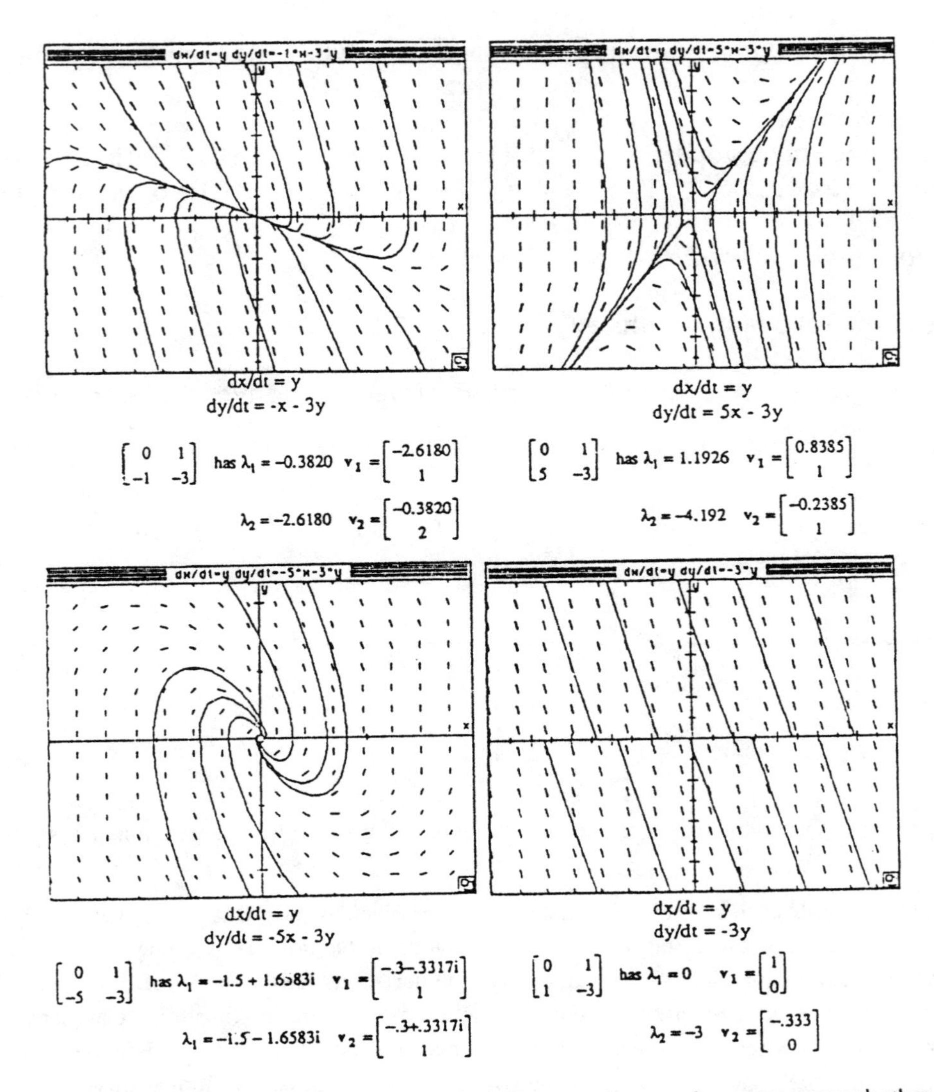

Figure 7: What does it *mean* for an ODE to be *linear*? (Instructor's Solution Sheet, [9])

There are lots of resources. One of my first and deepest favorites, with the biggest number of possibilities, is by Tony Danby at North Carolina State University. His first book is out of print, but he has just published a new one with even more examples. See [8].

References [5] and [6] are additional goldmines of modern examples. Others that are particularly good sources of materials are CODEE members, the Boston University group of Blanchard/Devaney/Hall, the Colvin/Hartig/McDill team at Cal Poly, David Lomen at University of Arizona, Herman Gollwitzer at Drexel, and John Polking at Rice.

References

Books and Journals

1. M. Artigue, V. Gautheron, *Systèmes Différentiels: Étude Graphique*, (CEDIC Paris 1983).
2. R. Borrelli, C. Coleman, *Differential Equations: A Modeling Perspective*, (Wiley 1997).
3. R. Borrelli, C. Coleman, W. Boyce, *Differential Equations Laboratory Workbook* (Wiley 1992).
4. M. Braun, C. Coleman, D. Drew, editors, *Differential Equations Models*, Volume I of *Modules in Applied Mathematics*, edited by W. Lucas (Springer-Verlag 1982).

5. CODEE Newsletter, by NSF Consortium for Ordinary Differential Equations Experiments. (Earlier issues can be viewed and downloaded at http://www.math.hmc.edu/codee/).
6. *College Mathematics Journal* (MAA), Vol. 25, November 1994. Special issue on innovation in teaching differential equations.
7. J. Countryman, *Writing to Learn Mathematics*, (Heinemann Press, 1992).
8. J.M.A. Danby, *Computer Modeling: from Sports to Spaceflight with Differential Equations: A Numerical Approach From Order to Chaos*, (Willmann-Bell, Inc., PO Box 35025, Richmond, VA 23235, 1997).
9. B. Felsager, B. West, *Laboratory Manual for Differential Equations* (Mathematics Department, Cornell University 1991).
10. J. Hubbard, B. West, *Differential Equations: A Dynamical Systems Approach, Part I, One Dimensional Systems*, (Springer-Verlag 1991, 1997).
11. J. Hubbard, B. West, *Differential Equations: A Dynamical Systems Approach, Part II, Higher Dimensional Systems*, (Springer-Verlag 1995).

Software

12. J. Hubbard, B. West, *MacMath*, (Springer-Verlag 1992, 1993).
13. B.West, S.Strogatz, J. McDill, J. Cantwell with H. Hohn, *Interactive Differential Equations*, (Addison Wesley Interactive 1996, 1997).
14. *ODE Architect* by CODEE Consortium (Wiley 1997).

For a longer list of software available, see the World Wide Web: http://www.math.hmc.edu/codee/